DOT/FAA/AR-06/44
DOT-VNTSC-FAA-06-21

Air Traffic Organization Operations Planning

Human Factors Research and Engineering Group

Washington, DC 20591

A Tool Kit for Evaluating Electronic Flight Bags

Divya C. Chandra

Michelle Yeh

U.S. Department of Transportation

Research and Innovative Technology Administration

John A. Volpe National Transportation Systems Center

Cambridge, MA 02142

September 2006

This document is available to the public through the National Technical Information Service, Springfield, Virginia, 22161

Notice

This document is disseminated under the sponsorship of the Department of Transportation in the interest of information exchange. The United States Government assumes no liability for its contents or use thereof.

Notice

The United States Government does not endorse products or manufacturers. Trade or manufacturers' names appear herein solely because they are considered essential to the objective of this report.

REPORT DOCUMENTATION PAGE			Form Approved
			OMB No. 0704-0188
Public reporting burden for this collection of information is estimated to average 1 hour per response, including the time for reviewing instructions, searching existing data sources, gathering and maintaining the data needed, and completing and reviewing the collection of information. Send comments regarding this burden estimate or any other aspect of this collection of information, including suggestions for reducing this burden, to Washington Headquarters Services, Directorate for Information Operations and Reports, 1215 Jefferson Davis Highway, Suite 1204, Arlington, VA 22202-4302, and to the Office of Management and Budget, Paperwork Reduction Project (0704-0188), Washington, DC 20503.			
1. AGENCY USE ONLY (Leave blank)	2. REPORT DATE September 2006		3. REPORT TYPE AND DATES COVERED
4. TITLE AND SUBTITLE A Tool Kit for Evaluating Electronic Flight Bags			5. FUNDING NUMBERS FA6Y CD304
6. AUTHOR(S) Divya C. Chandra and Michelle Yeh			
7. PERFORMING ORGANIZATION NAME(S) AND ADDRESS(ES) U.S. Department of Transportation John A. Volpe National Transportation Systems Center Research and Innovative Technology Administration Cambridge, MA 02142-1093			8. PERFORMING ORGANIZATION REPORT NUMBER DOT-VNTSC-FAA-06-21
9. SPONSORING/MONITORING AGENCY NAME(S) AND ADDRESS(ES) U.S. Department of Transportation Federal Aviation Administration Air Traffic Organization Operations Planning Human Factors Research and Engineering Group 800 Independence Avenue, SW Washington, D.C. 20591 Program Manager: Dr. Tom McCloy			10. SPONSORING/MONITORING AGENCY REPORT NUMBER DOT/FAA/AR-06/44
11. SUPPLEMENTARY NOTES			
12a. DISTRIBUTION/AVAILABILITY STATEMENT This document is available to the public through the National Technical Information Service, Springfield, VA 22161			12b. DISTRIBUTION CODE
13. ABSTRACT (Maximum 200 words) Over the past few years, the Volpe Center has developed a set of five tools that can be used to evaluate Electronic Flight Bags (EFBs) from a human factors perspective. The goal of these tools is to help streamline and standardize EFB human factors assessments by the Federal Aviation Administration (FAA). This document introduces all of the Volpe EFB assessment tools, which are for use by any FAA or industry evaluator. This report contains descriptions of the tools and practical information on when and how to use each tool. It is not necessary to be a human factors expert to use the tools, and the tools can be incorporated into the evaluation process in different ways depending on the time available. The appendices to this report contain the full version of the tools.			
14. SUBJECT TERM Electronic Flight Bag, EFB, design approval, operational approval, usability, evaluation, human factors tool			15. NUMBER OF PAGES 64
			16. PRICE CODE
17. SECURITY CLASSIFICATION OF REPORT Unclassified	18. SECURITY CLASSIFICATION OF THIS PAGE Unclassified	19. SECURITY CLASSIFICATION OF ABSTRACT Unclassified	20. LIMITATION OF ABSTRACT

This page left blank intentionally.

Preface

This report was prepared by the Human Factors Division of the Office of Aviation Programs at the John A. Volpe National Transportation Systems Center. It was completed with funding from the FAA Human Factors Research and Engineering Group (AJP-61) in support of the Aircraft Certification Service Avionics Branch (AIR-130) and the Flight Standards Service Flight Technologies and Procedures Division (AFS-430). We would like to thank our FAA program manager, Tom McCloy, as well as our technical sponsors, Rich Adams, Bill Kaliardos, Colleen Donovan, and the many other FAA staff who have given us feedback and suggestions.

The views expressed herein are those of the authors and do not necessarily reflect the views of the Volpe National Transportation Systems Center, the Research and Innovative Technology Administration, or the United States Department of Transportation.

Feedback on this document can be sent to Divya Chandra (Divya.Chandra@volpe.dot.gov) or Michelle Yeh (Michelle.Yeh@volpe.dot.gov). Further information on this research effort, including reports on how the tools were developed and tested, can be found at http://www.volpe.dot.gov, under the Human Factors Division, Electronic Flight Bag section.

METRIC/ENGLISH CONVERSION FACTORS

ENGLISH TO METRIC

LENGTH (APPROXIMATE)
- 1 inch (in) = 2.5 centimeters (cm)
- 1 foot (ft) = 30 centimeters (cm)
- 1 yard (yd) = 0.9 meter (m)
- 1 mile (mi) = 1.6 kilometers (km)

AREA (APPROXIMATE)
- 1 square inch (sq in, in^2) = 6.5 square centimeters (cm^2)
- 1 square foot (sq ft, ft^2) = 0.09 square meter (m^2)
- 1 square yard (sq yd, yd^2) = 0.8 square meter (m^2)
- 1 square mile (sq mi, mi^2) = 2.6 square kilometers (km^2)
- 1 acre = 0.4 hectare (he) = 4,000 square meters (m^2)

MASS - WEIGHT (APPROXIMATE)
- 1 ounce (oz) = 28 grams (gm)
- 1 pound (lb) = 0.45 kilogram (kg)
- 1 short ton = 2,000 pounds (lb) = 0.9 tonne (t)

VOLUME (APPROXIMATE)
- 1 teaspoon (tsp) = 5 milliliters (ml)
- 1 tablespoon (tbsp) = 15 milliliters (ml)
- 1 fluid ounce (fl oz) = 30 milliliters (ml)
- 1 cup (c) = 0.24 liter (l)
- 1 pint (pt) = 0.47 liter (l)
- 1 quart (qt) = 0.96 liter (l)
- 1 gallon (gal) = 3.8 liters (l)
- 1 cubic foot (cu ft, ft^3) = 0.03 cubic meter (m^3)
- 1 cubic yard (cu yd, yd^3) = 0.76 cubic meter (m^3)

TEMPERATURE (EXACT)
[(x-32)(5/9)] °F = y °C

METRIC TO ENGLISH

LENGTH (APPROXIMATE)
- 1 millimeter (mm) = 0.04 inch (in)
- 1 centimeter (cm) = 0.4 inch (in)
- 1 meter (m) = 3.3 feet (ft)
- 1 meter (m) = 1.1 yards (yd)
- 1 kilometer (km) = 0.6 mile (mi)

AREA (APPROXIMATE)
- 1 square centimeter (cm^2) = 0.16 square inch (sq in, in^2)
- 1 square meter (m^2) = 1.2 square yards (sq yd, yd^2)
- 1 square kilometer (km^2) = 0.4 square mile (sq mi, mi^2)
- 10,000 square meters (m^2) = 1 hectare (ha) = 2.5 acres

MASS - WEIGHT (APPROXIMATE)
- 1 gram (gm) = 0.036 ounce (oz)
- 1 kilogram (kg) = 2.2 pounds (lb)
- 1 tonne (t) = 1,000 kilograms (kg) = 1.1 short tons

VOLUME (APPROXIMATE)
- 1 milliliter (ml) = 0.03 fluid ounce (fl oz)
- 1 liter (l) = 2.1 pints (pt)
- 1 liter (l) = 1.06 quarts (qt)
- 1 liter (l) = 0.26 gallon (gal)
- 1 cubic meter (m^3) = 36 cubic feet (cu ft, ft^3)
- 1 cubic meter (m^3) = 1.3 cubic yards (cu yd, yd^3)

TEMPERATURE (EXACT)
[(9/5) y + 32] °C = x °F

QUICK INCH - CENTIMETER LENGTH CONVERSION

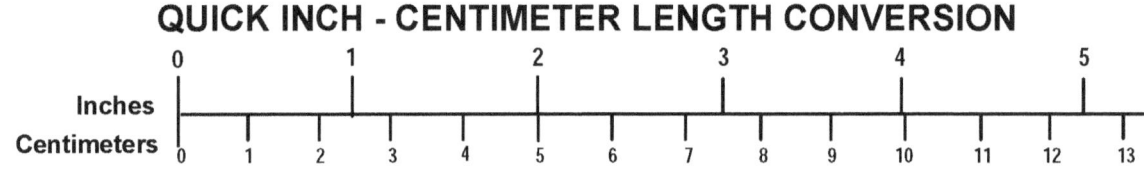

QUICK FAHRENHEIT - CELSIUS TEMPERATURE CONVERSION

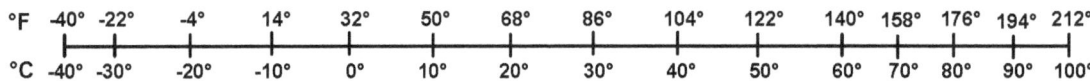

For more exact and or other conversion factors, see NIST Miscellaneous Publication 286, Units of Weights and Measures. Price $2.50
SD Catalog No. C13 10286   Updated 6/17/98

Table of Contents

Preface v
Table of Contents vii
List of Figures viii
List of Tables viii
Executive Summary ix
1. Introduction 1
2. Background 1
3. Overview of EFB Assessment Tools 2
4. Using the EFB Assessment Tools 5
5. Summary and Conclusions 5
6. References 6
Appendix A: EFB Human Factors Design Review Checklist A.1
Appendix B: EFB User-Interface Assessment Tool B.1
Appendix C: Guide for Developing Simulator and Validation Flight Scenarios C.1
Appendix D: Operational Evaluation Questions D.1
Appendix E: Line Operations Evaluation Job Aid E.1
Appendix F: Tool Information Tables F.1

List of Figures

Figure 3-1 Overview of EFB Usability Assessment Tools. 2

List of Tables

Table 3-1. Comparison of EFB usability assessment tools. 4
Table F-1. EFB Human Factors Design Review Checklist..F.1
Table F-2. EFB User-Interface Assessment Tool F.2
Table F-3. Guide for Developing Simulator and Validation Flight Scenarios......................................F.3
Table F-4. Operational Evaluation Questions..F.4
Table F-5. Line-Operations Evaluation Job Aid F.5

Executive Summary

Over the past few years, the Volpe Center has developed a set of five tools that can be used to evaluate Electronic Flight Bags (EFBs) from a human factors perspective. The goal of these tools is to help streamline and standardize EFB human factors assessments by the Federal Aviation Administration (FAA). The tools were developed and documented for the FAA in order to facilitate the identification and resolution of human factors/pilot interface issues with EFB systems, either in terms of design and/or operational use. The tools are designed for use by evaluators who are not human factors experts. They can be used at different stages of EFB development for different types of evaluations.

By understanding the tools before beginning the approval process, the most appropriate tools can be selected, customized, and incorporated into the EFB evaluation at relatively little incremental cost. This report contains descriptions of the tools and practical information on when and how to use each tool. The appendices to this report contain the full version of every tool. Specifically, the five tools discussed are:

1) EFB Human Factors Design Review Checklist (Appendix A)
2) EFB User-Interface Assessment Tool (Appendix B)
3) Guide for Developing Simulator and Validation Flight Scenarios (Appendix C)
4) Operational Evaluation Questions (Appendix D)
5) Line-Operations Evaluation Job Aid (Appendix E)

While the tools were designed for use by regulatory authorities, they are also of use to industry manufacturers and customers who could use them during their design/evaluation process to improve the system or to anticipate the results of a regulatory evaluation. The biggest benefit of using these tools, however, is that their early use may reduce the need for costly redesigns associated with poor system interfaces, and ensure that the EFB system is more usable in the long run, which produces benefits for everyone—the regulatory authority, the manufacturer, the customer, and the pilot.

This page left blank intentionally.

1. Introduction

Over the past few years, the Volpe Center has developed a set of five tools that can be used to evaluate Electronic Flight Bags (EFBs) from a human factors perspective. All of these tools were developed at the request of the Federal Aviation Administration (FAA) in order to facilitate the identification and resolution of human factors/pilot interface issues with EFB systems, either in terms of design and/or operational use. The tools are needed because EFBs are sophisticated devices that may be approved for use through a relatively quick process, in accordance with the guidance in the 2003 FAA Advisory Circular (AC) 120-76A, *Guidelines for the certification, airworthiness, and operational approval of electronic flight bag computing devices* (Federal Aviation Administration [FAA], 2003). It is assumed that the reader is familiar with AC 120-76A and the related draft EFB Job Aid (FAA, 2006).

The goal of the EFB usability assessment tools is to help streamline and standardize EFB human factors assessments by the FAA. We expect that the tools will benefit the FAA, system designers, and operators by providing structure for human-factors evaluations. The tools ensure that all parties are well informed about the evaluation, and facilitate consistent documentation for the approval. In addition, the tools are available for use by industry, to help them improve their EFB systems and to anticipate the results of a regulatory evaluation.

This report contains descriptions of the tools and practical information on when and how to use each tool. It was written with the FAA evaluator in mind, but may be used by any evaluator. The evaluator might be an FAA inspector, an EFB designer/developer, or even an EFB customer/operator. Evaluators who use these tools are not expected to be human factors experts.

The full version of each of the five tools is included in the appendices to this report. These tools can be incorporated into the evaluation process in different ways, depending on the time available. Even a short evaluation (less than one hour) can provide valuable information about how the EFB system will function in the flight deck. For background information on how the tools were developed and tested, see Chandra and Yeh (2006), Chandra, Yeh, and Riley (2004), Chandra and Yeh (2004), and Chandra (2003). Another important resource document is the report, *Human Factors Considerations in the Design and Evaluation of Electronic Flight Bags (EFBs)*, by Chandra, Yeh, Riley, and Mangold (2003), which contains detailed information on EFB human factors considerations.

2. Background

The Federal Aviation Administration (FAA) is seeing an increasing number of applications for EFB approvals. The main FAA EFB approval guidance is contained in AC 120-76A (FAA, 2003). The draft EFB Job Aid initiated by FAA Flight Standards (AFS-400) provides additional clarification for field inspectors (FAA, 2006). The draft EFB Job Aid is intended to supplement and clarify the recommendations and processes outlined in AC 120-76A. It is not intended to replace or supersede AC 120-76A.

The guidance in AC 120-76A is complex. One reason for this complexity is that the EFB AC covers both the equipment issues typically addressed by the engineers and test pilots in the FAA Aircraft Certification Service, as well as operational approval and procedures issues typically addressed by the FAA's Flight Standards inspectors and Aircraft Evaluation Group (AEG) pilots. All EFBs that are approved under the EFB AC require an *operational evaluation* to ensure that the flight crew can *use* the new system safely (e.g., without undue distraction, and with appropriate training and procedures). In particular, complex applications that require crew interaction, such as flight performance calculators or electronic charts, will undergo a more formal review than simpler, non-interactive applications, such as electronic document viewers. Systems that have more complex hardware will also require a *design approval* to ensure that the EFB does not impair the functionality of existing flight deck systems.

In addition to the FAA policy and guidance documents noted above, the Volpe Center has produced a number of research reports on EFB human factors research issues. An early key document was a Volpe Center report on human factors considerations for the design and evaluation of EFBs that is referenced within the EFB AC (Chandra, et al., 2003). This report is a comprehensive technical reference that contains detailed supporting material on EFB design and evaluation. While it could be used in practice as an aid for the evaluation of EFBs, there was a need to develop shorter, more focused tools for that purpose. Therefore, the Volpe Center developed two tools to aid Aircraft Certification in their EFB evaluations. The process of developing these tools is documented in Chandra et al. (2004). Subsequently, the Volpe Center developed three tools specifically for Flight Standards operational approvals. These three tools are incorporated as appendices in the draft EFB Job Aid (FAA, 2006).

3. Overview of EFB Assessment Tools

This report describes five EFB human factors assessment tools developed by the Volpe Center specifically to be aids in conducting practical and focused EFB evaluations. These tools are listed below and are provided in full as appendices at the end of this report:

1) EFB Human Factors Design Review Checklist (Appendix A)
2) EFB User-Interface Assessment Tool (Appendix B)
3) Guide for Developing Simulator and Validation Flight Scenarios (Appendix C)
4) Operational Evaluation Questions (Appendix D)
5) Line-Operations Evaluation Job Aid (Appendix E)

Note that there is no requirement for either the FAA or industry to use any of these tools.

Figure 1 shows an overview of the tools from the perspective of when they can be used, relative to the maturity of the EFB system. The solid gray lines in Figure 1 mark the best time for using each tool, and the dashed gray lines denote other periods when the tools could be useful. The EFB Human Factors Design Checklist and the EFB User-Interface Assessment Tool focus on aspects of the EFB hardware and software only, while the three other tools consider the system as a whole, and consider pilot training and procedures as well.

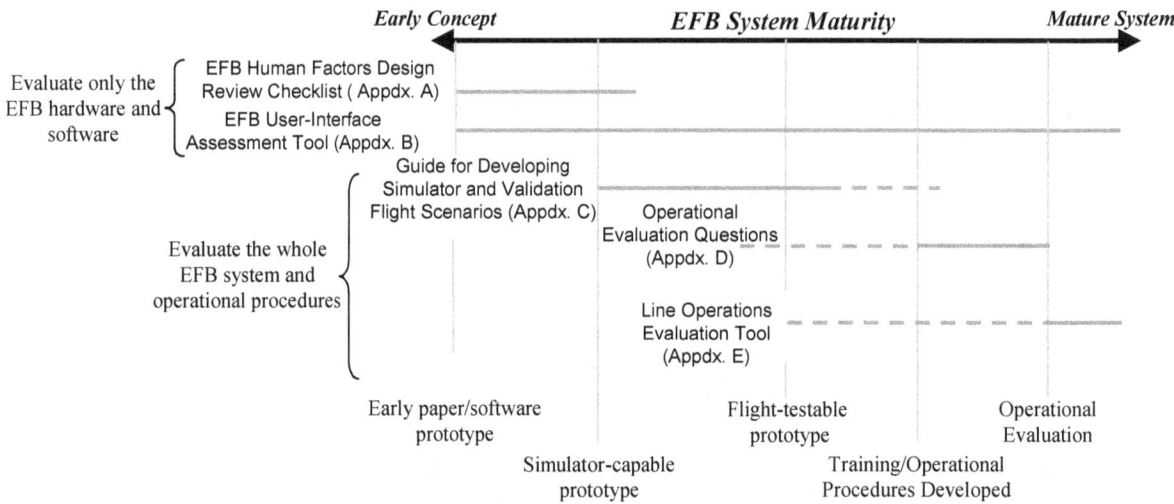

Figure 3-1 Overview of EFB Usability Assessment Tools.

The EFB Human Factors Design Checklist (Appendix A) is most appropriate for evaluating a system in the early stages of design and development because it focuses on specific design aspects, such as selection of fonts. In contrast, the EFB User-Interface Assessment Tool (Appendix B) is a more versatile tool. It can be used at any time during system development, even into evaluations of line operations. The EFB User-Interface Assessment Tool is designed to identify the significant interface issues from the users' perspective, e.g., overall consistency. Issues identified by this tool are ones that may have been originally overlooked by designers. Both of these tools are based on the foundation provided in the Volpe Center report on EFB human factors considerations (Chandra, et al., 2003).

The three tools in Figure 1 that consider the operational use of the EFB may be applied at different stages of system maturity. The Guide for Developing Simulator and Validation Flight Scenarios (Appendix C) helps the evaluator develop and run simulator scenarios that are too costly or too risky to test in real flights. This guide can be used prior to the development of a flight-testable unit. As the system matures, this guide may also be helpful in developing and testing operational procedures for using the EFB, especially for an air transport environment, as indicated by the dashed line extending to the right.

Simulator and/or in-flight validation tests may be needed to fully determine the suitability of an EFB (see AC 120-76A Paragraph 12 (j), pp. 21-22). These tests may be done for a number of reasons to ensure compliance with the regulations. As mentioned earlier, these tests typically consider both the EFB system and the operational context in which it is used. Some of the suggested simulated emergency procedures may only be appropriate in a simulator or training device. While these tests may be expensive, they do serve a unique and important function, and may be necessary in some cases.

The Operational Evaluation Questions (Appendix D) address the user interface from both an operational and design perspective. The questions can be used prior to developing a flight-testable unit, but its utility may be limited until a more mature system is developed. Some questions in this set may be useful to the manufacturer or customer earlier in development, as shown by the dashed line in Figure 1, but the FAA inspector is most likely to use the Operational Evaluation Questions only after training and operational procedures for using EFBs have been developed by the operator.

The Line Operations Evaluation Tool (Appendix E) contains questions that address the impact of the EFB on safety and operations. It is most appropriate for use during a line operational evaluation, i.e., when the system has been deployed and it being observed and evaluated during initial use. This would be the first time that line crews use the system, so it is a good stage to assess whether the EFB system can be used by the "average" crew, as opposed to those who have a high level of system knowledge, and perhaps a special interest in its success.

Table 1 contains a comparison of the five different Volpe Center assessment tools discussed in this report. Each tool is compared along five dimensions: scope, user(s), investment, benefit, and limitations. Scope refers to the characteristics that are addressed in the evaluation and the environment in which the evaluation is conducted. The User(s) column identifies characteristics of the evaluators who could use the tool. The Investment column indicates how much time and other resources would be required to conduct an evaluation with a particular tool. The Benefits column describes what types of results one can expect. The Limitations column lists any caveats on using the tool, e.g., issues that will not be addressed.

Tables F-1 through F-5 in Appendix F provide more information on the tools in terms of how they can be used and what types of findings an evaluator can expect. In particular, the Appendix F tables provide a more in-depth description of the tool, and they consider the process for using the tool, as well as the resulting documentation. Each table in the Appendix F contains the following information: Scope and Description, User, Process, Documentation, Time and Resources, Benefits, and Limitations.

Tool	Scope	User(s)	Investment	Benefits	Limitations
EFB Human Factors Design Review Checklist	Analytical ("desk-top") detailed assessment of user-interface design • Specific items for some common applications	Best suited for applications developers; designed for Aircraft Certification • Any level of human-factors expertise	Low to moderate • Office environment • Approximately one day for evaluation; half-day for the simplest EFB systems	• Uncovers specific design issues (e.g., font choice) quickly	• Best used relatively early in the system development • Does not address operational use of system (e.g., training/procedures)
EFB User-Interface Assessment Tool	Analytical ("desk-top") high-level assessment of user-interface design • Specific items for some common applications	Broad Range; designed for Aircraft Certification • Operator • FAA Inspector • Applications developers • Any level of human-factors expertise	Very low to moderate • Office environment • Short time duration up front (e.g., 1 hour) • Optional additional time for data synthesis (a few days)	• Uncovers "big" issues (e.g., potential for confusion) quickly • With data synthesis, can uncover subtle structural problems • Good for validating EFB system design concept	• Data are subjective and qualitative so the impact of the issues is difficult to document • Does not address operational use of system (e.g., training/procedures)
Guide for Developing Simulator and Validation Flight Scenarios	EFB system design (both installation and user-interface), and operational use of the EFB system, especially in unusual operating conditions • Note that the tool provides examples, but needs to be heavily tailored	Best suited for aircraft manufacturers or operators because of need for simulator or aircraft for the tests; designed for Flight Standards • FAA may request a simulator or validation flight during the approval process • Human factors expertise is beneficial	High, but may help to identify issues during simulator tests as opposed to more costly flight tests • May be worth the cost for testing sophisticated, or highly complex EFB systems (e.g., EFBs that are integrated with aircraft systems)	• Validates overall system use under unusual operating conditions (e.g., low-visibility operations) • Could be used throughout EFB system development (from concept to mature design) • Could provide quantitative data	• May not be worth the cost for simple or evolutionary EFB systems • Data analysis could be complex; human factors expertise may be required
Operational Evaluation Questions	Comprehensive analytic ("desk-top") assessment of EFB system operational use and system design • Intended for use in approving initial EFB system installations	Intended user is the Flight Standards inspector from the Aircraft Evaluation Group • Operator and manufacturer should be prepared to support the FAA inspector's evaluation • Requires experienced evaluators (not necessarily human factors expertise)	Moderate • Approximately one day for evaluation • Some parts can be done in a desk-top environment, but the aircraft and installation should be well understood	• Validates initial use of EFB system for a particular flight deck • Considers all aspects of EFB use, including system design, installation, training, and procedures	• Relies upon evaluator's experience in customizing the tool for the evaluation and in making appropriate assessments
Line Operations Evaluation Job Aid	Practical ("line-operations") assessment of EFB system operational use • Intended for use in evaluating EFB system use over multiple observations	Best for operator and FAA Flight Standards operations inspector • Any level of human-factors expertise • Operators could customize the tool to provide more quantitative data for internal assessments	Very low to moderate • Use to record notes during and after observation flights • Collect records for multiple flights. Data could be aggregated and analyzed across the observations	• Validates overall system use in normal operations at a relatively low cost • Can uncover training/procedural issues • Can uncover variances between end-users (pilots) via multiple flight observations	• Does not address design of EFB system • In simplest form, does not collect quantitative data

Table 3-1. Comparison of EFB usability assessment tools.

4. Using the EFB Assessment Tools

The evaluator should review all the tools at the beginning of the EFB approval process to determine which of the tools can be used effectively for that particular application, and how the tools could be incorporated into the normal process. The evaluator should keep in mind several factors, including:

1) Overall goal for the evaluation. What aspects of the EFB are being evaluated?
2) System maturity at the time of the review
3) Time available per review
4) Number of opportunities for reviewing the system during the approval process
5) Number of evaluators who will participate in the reviews

For example, in some cases, the evaluator may see the EFB multiple times throughout system development. In other cases, the evaluator may only see the system once, just as it is ready to be deployed. Sometimes a team will evaluate the system, and other times an individual will be responsible. After the evaluation parameters are specified, the evaluator can select the tools.

The EFB User-Interface Assessment tool ([Appendix B](#)) was tailored for Aircraft Certification Specialists who typically focus on the EFB design and airworthiness aspects of the evaluation and approval. The primary tools for Flight Standards operational approvals will be those that consider training and operating procedures (Appendices C, D, and E). System developers and/or operators may choose to use the tools more comprehensively, combining them for increased understanding of the issues. For example, if a simulator or flight evaluation is constructed with the aid of the Guide for Developing Simulator and Validation Flight Scenarios, the EFB User-Interface Assessment Tool, and even the Line Operations Evaluation Tool could be used within the tests to collect data. Also, because system developers and operators will review the system as it matures, they have the best opportunity to track progress on issues using the tools across multiple evaluation stages.

Once the assessment tools have been selected, the next step is to customize the tools for the specific situation. While the Volpe EFB assessment tools provide a good starting point, they need to be tailored for the specific situation because EFB systems vary widely. Some systems are designed for airline operators with numerous crews; other systems are designed for small operators with just a few aircraft and crews, who operate under different FAA regulations than the airlines. The EFB systems may be relatively simple stand-alone devices, or they could be installed in the aircraft (e.g., the Boeing 777 EFB) (Yeh and Chandra, 2005). The tools do not make any assumptions about the capabilities of the EFB, or the complexity of their use in operations, but the inspector needs to takes these topics into account when customizing the tools.

In a regulatory situation, the FAA evaluator should discuss use of the tools with the applicant as early as possible, to ensure that everyone understands how the tools will be incorporated into the evaluation process and how their results will be used. Note that there is no "wrong" way to use the tools. The tools are aids that help promote a thoughtful structured evaluation of the EFB. With the exception of a few items typically related to regulatory guidance or potential safety concerns, the tools do not provide the evaluator the "best" or "correct" way to design the EFB. Any issues that are uncovered during the human factors evaluation should be resolved together between the FAA and applicant. The final assessment and plan for action should be based on the evaluator's best judgment and the opinions of the applicant.

5. Summary and Conclusions

Five different tools for assessing EFB from a human factors perspective are described in this report. These tools can be used at different stages of EFB development for different types of evaluations. By understanding the tools before beginning an EFB approval process, the most appropriate tools can be selected, customized, and incorporated into the EFB evaluation at relatively little incremental cost.

The tools have been developed and documented for the FAA to facilitate the identification and resolution of human factors/pilot interface issues with the EFB system, either in terms of design and/or operational use. The tools are also of use to the EFB manufacturers and customers, who could use the tools to improve their system design or to anticipate the results of a regulatory evaluation. The biggest benefit of using these tools, however, is that their early use can reduce the redesign associated with poor system interfaces, and ensure that the EFB system is more usable in the long run, which produces benefits for everyone—the regulatory authority, the manufacturer, the customer, and the pilot.

6. References

Key Documents and Reports

Chandra, D. C., Yeh M., Riley, V., & Mangold, S.J. (2003). Human factors considerations in the design and evaluation of Electronic Flight Bags (EFBs), Version 2. DOT-VNTSC-FAA-03-07. USDOT Volpe Center: Cambridge, MA. Available at http://www.volpe.dot.gov/opsad/efb and http://www.hf.faa.gov, under Library Documents.

Federal Aviation Administration, Advisory Circular (AC) 120-76A, March 17, 2003. *Guidelines for the certification, airworthiness, and operational approval of electronic flight bag computing devices.* Available at www.faa.gov, under Regulations and Policies.

Federal Aviation Administration, *Electronic Flight Bag (EFB) Job Aid, Version 1.0.* Draft of 26 January 2006. Available at http://www.volpe.dot.gov/opsad/efb and www.faa.gov, under Flight Standards Service, Flight Technologies and Procedures Division, Flight Technologies Requirements Branch.

Yeh, M. and Chandra, D. C. (2005) Electronic Flight Bag (EFB): 2005 Industry Review. Report No. DOT-VNTSC-FAA-05-06. Cambridge, MA: USDOT Volpe Center. Available at http://www.volpe.dot.gov/opsad/efb

Supplementary Documents

Chandra, D.C. and Yeh, M. (2006) Evaluating Electronic Flight Bags in the Real World. *Proceedings of the International Conference on Human-Computer Interaction in Aeronautics (HCI–Aero) 2006.* 20–22 September 2006, Seattle, Washington. Available at http://www.volpe.dot.gov/opsad/efb

Chandra, D.C., Yeh, M. and Riley, V. (2004) *Designing a Tool to Assess the Usability of Electronic Flight Bags (EFBs).* DOT/FAA/AR-04/38 and DOT-VNTSC-FAA-04-12. Cambridge, MA: USDOT Volpe Center. Available at http://www.volpe.dot.gov/opsad/efb and http://www.hf.faa.gov, under Library Documents.

Chandra, D.C. and Yeh, M. (2004). Designing and Testing a Tool for Evaluating Electronic Flight Bags. *Proceedings of the International Conference on Human-Computer Interaction in Aeronautics (HCI–Aero) 2004.* 29 September – 1 October 2004, Toulouse, France. Available at http://www.volpe.dot.gov/opsad/efb

Chandra, D.C. (2003). A tool for structured evaluation of electronic flight bag usability. In *Proceedings of the 22nd Digital Avionics Systems Conference (DASC).* 12–16 October 2003, Indianapolis, IN. Available at http://www.volpe.dot.gov/opsad/efb

Appendix A: EFB Human Factors Design Review Checklist

2 General EFB System
HARDWARE CONSIDERATIONS

2.1.5 Legibility—Lighting Issues
- ❖ Automatic brightness adjustment should be independent for each EFB (See AC 25-11)
- ❏ Screen brightness should adjustable in fine increments or continuously
- ❏ Buttons and labels should be adequately illuminated for night use

2.2.4 Kneeboard EFBs
- ❖ Kneeboard EFB should be easily removable

2.4.1 User Interface—General Design
- ❏ User interface should have a consistent set of controls and graphical elements (see also General Principles)
- ❏ Controls used for different functions should be visually distinct
- ❏ Graphic elements and controls should follow personal computer conventions, except where clearly inappropriate for flight deck environment (see also General Principles)

2.5.1 Pointing and Cursor Control Devices
- ❏ Input devices should be selected and customized based on the type and complexity of the entries to be made and flight deck environmental factors that affect its usability
- ❏ Performance parameters should be tailored for the intended application and for the flight deck environment
- ❏ Users should be able to rest and/or stabilize their hand when using the pointer or cursor control device
- ❏ Active areas should be sized to permit accurate selection with the pointer/cursor device under all operating conditions

2.5.2 Hardware Controls
- ❖ All controls should be properly labeled (14 CFR 23.1555, 25.1555, and 27.1555)
- ❖ All soft function keys should be labeled
- ❖ Inactive soft function keys should not be labeled or should use a visual convention to indicate that the function is not available
- ❏ Physical function keys should provide tactile feedback when pushed
- ❏ Key repeats should be filtered by the software if they occur too closely together
- ❏ Soft function keys should be drawn in a reserved space outside the main content area
- ❏ The same function should appear on the same function key, whenever possible
- ❏ Labels should be consistent
- ❏ Labels should be clear and brief
- ❏ Labels may use standard abbreviations; ambiguous abbreviations should be avoided
- ❏ Labels should be located near the controls they identify and should not be confusingly close to other labels or other controls
- ❏ Labels should be drawn in horizontal text
- ❏ Physical controls should be collocated with the display
- ❏ The most frequently used controls should be placed at the most accessible locations
- ❏ Controls presented in a small space may need to be grouped according to function and/or order of use
- ❏ Controls should be designed to deter inadvertent activation

2.5.3 Display
- ☐ The physical nature of the display screen should minimize the likelihood that information will be obscured

2.5.5 Keyboards
- ☐ Keyboard type should be appropriate for the given task
  - ▪ QWERTY type keyboards should be used for text entry
  - ▪ Numeric keypads are best suited for significant numeric entries
- ☐ Keyboards should provide appropriate tactile feedback
- ☐ Users should be able to rest/stabilize their hand to use the keyboard, especially during turbulence

SOFTWARE
Symbols and Graphical Icons

2.4.4 Graphical Icons
- ☐ Icons should be accompanied with text labels
- ☐ Design of icons should minimize training and maximize intuitiveness for cross-cultural use

2.4.13 Non-Text Display Elements
- ☐ Non-text display elements should be distinguishable based on shape alone, without relying on secondary cues such as color or labels
- ☐ Non-text display elements should be designed for legibility on minimum expected display resolution viewed from the maximal intended viewing distance

Formatting/Layout

2.4.10 Legibility of Text—Characters
- ☐ Typeface should be highly legible. HFDS recommends:
  - Spare use of upper case text (8.2.5.8.2)
  - Mixed upper and lower case for continuous text (8.2.5.8.4)
  - Serif fonts for high resolution displays (8.2.5.7.5)
  - Sans serif fonts otherwise (8.2.5.7.6)
  - Character contrast between 6:1 and 10:1 (8.2.5.6.12)
  - Characters stroke width 10 to 12% of character height (8.2.5.6.14)
- ☐ Individual characters should not be easily confused with other characters
- ☐ Slanting or italic text should be avoided

2.4.11 Legibility of Text—Typeface Size and Width
- ❖ Typeface should be appropriate for viewing distance, lighting conditions, and text criticality
- ☐ The FAA HFDS recommends that:
  i. Minimum character height should be 1/200 of viewing distance, e.g., for 35" viewing distance, 0.175" tall (17.5 pixels at 100 pix/inch) (8.2.5.6.6)
  ii. Preferred character height should be 1/167 of viewing distance (8.2.5.6.5)
  iii. Character height to width ratios should be (8.2.5.6.10)
      - o <80 char per line, 1 to 0.7 up to 0.9 (15 pix tall, 10.5 to 13.5 pix wide) for monotype fonts
      - o >80 char per line, at least 1 to 0.5 (15 pix tall, 7.5 pix wide)
      - o 1:1 for M and W in a proportional font
- ☐ Larger fonts should be used for text read in poor viewing conditions

2.4.12 Legibility of Text—Spacing for Readability
- ❏ Text should be spaced appropriately to facilitate reading
- ❏ Line lengths should be appropriate for text content
- ❏ To facilitate readability, HFDS recommends the following:
  (a) Use horizontal spacing between characters that is at least 10% of character height (15 pix tall, 1.5 pix spacing) (8.2.5.6.1)
  (b) Use spacing between words of at least one character for equally spaced characters, or width of "N" for proportional fonts (8.2.5.6.2)
  (c) Use spacing between lines of at least two stroke widths or 0.15 of character height (15 pix tall, 2.25 pix leading), whichever is greater (8.2.5.6.3)
  (d) Separate paragraphs with blank line (8.2.5.6.4)

Interactions: Accessing Functions and Options

2.4.5 Multi-Tasking
- ❏ The user should be able to identify the active application easily
- ❏ The user should be able to:
   - Select which of the open applications is currently active
   - Switch between applications easily
- ❏ Applications, running in the background, should be in the same state when the user returns to it, other than the completion of any background processing
- ❏ Responsiveness of an individual application should not suffer when all applications are running simultaneously
- ❏ The user should be able to exit applications with pending activities by completing them or by acknowledging that they are incomplete
- ❏ The system should discourage use of non-flight-related applications and ask for an extra confirmation to launch

2.4.6 Responsiveness
- ❖ The system should provide feedback when a user input is processed
   - Alphanumeric inputs should be shown within 0.2 seconds (SAE ARP 4791)
- ❖ A "system busy" indicator should be displayed if user inputs can not be processed within 0.5 seconds (SAE ARP 4791)
- ❏ The EFB applications should have a "system busy" indicator
- ❏ The type of feedback should be appropriate for the type of user input
- ❏ If tasks take more than a few seconds to complete, indicators should show their progress
- ❏ User entries made while the system is busy should be stored for later processing

2.4.7 Anchor Locations
- ❏ If the EFB supports more than one application, there should be an anchor location from which the user moves between applications
- ❏ Each EFB application should have its own anchor page
- ❏ It should be easy to move from any location in the EFB to an anchor location, and vice versa

2.4.18 Links to Related Material
- ❏ A consistent philosophy should be used for accessing different types of information. Similar types of information should be accessed in the same way
- ❏ Users should be able to keep track of how to move between topics. Users should be able to return to the starting point easily

Error Handling and Prevention

2.1.7   Failure Modes
- ❏ EFB should alert the flight crew to probable application/system failures (AC 120-76A, Section 10.e (2))

2.4.8   Display of System Status
- ❏ Any full or partial application failure should be indicated with a positive indicator (AC 120-76A, Section 10.d (2))
- ❏ The immediacy of indicator should be appropriate to the function that is lost or disabled (AC 120-76A, Section 10.d (2))

2.4.15   Ensuring Integrity of EFB Data
- ❖ EFB data should be checked prior to installation to ensure that they are accurate, current, and uncorrupted
- ❏ The EFB should check that the current date is within the valid date range
- ❏ The EFB should allow data with an effective date in the future to be installed
- ❏ The system should conduct a self-test to ensure that the data is current and generate a message to the flight crew if any data is out of date. The message should indicate where to go for further information.

2.4.17   Crew Confirmation of EFB Software/Database Approval
- ❖ The latest revision information should be available upon request

2.4.19   User-Interface Customization
- ❖ There should be an easy means to return all settings to their default values
- ❏ For Part 121 and 135, the default settings should be customizable only by an administrator
- ❏ For Part 91, the default settings should be specified by the manufacturer and configurable by the user

Multiple Applications

2.4.2   Application Compatibility and Style Guides
- ❏ All applications should follow a common style guide, preferably specific to that aircraft
- ❏ Color and other formatting should be internally consistent across applications (AC 120-76A, Section 10.b (1))
- ❏ Help facility, if available, should be standardized across applications
- ❏ Soft key labels and menus should be consistent across applications
- ❏ Common actions allowed on multiple applications should be performed in the same manner (see also Interactions: Accessing functions and options)
- ❏ Manufacturers should prepare style guides for third party developers

General Principles

2.4.1   User Interface—General Design
- ❏ User interface should have a consistent set of controls and graphical elements (see also Hardware)
- ❏ Graphic elements and controls should follow personal computer conventions, except where clearly inappropriate for flight deck environment (see also Hardware)
- ❏ Menu functions should be accessible in proportion to frequency of use and criticality to mission

2.4.3   General Use of Colors
- ❖ Red and amber should be reserved for highlighting *warning* and *caution* level conditions respectively (AC 120-76A, 10.d (1))
- ❖ Color should not be sole means of coding important differences in information; color should be used redundantly
- ❖ Color-coding scheme should be interpretable easily and accurately.
- ❏ Each color should be associated with only one meaning
- ❏ No more than six colors with assigned meanings should be used in a color-coding scheme

EFB Human Factors Design Review Checklist

- ❏ EFB colors should not conflict with flight deck conventions
- ❏ For Part 121 and 135, default colors that represent different types of data should be customizable only by an appropriately authorized administrator
- ❏ If colors are customizable, there should be an easy way to return to default settings

2.4.8   Alerts and Reminders

- ❖ Alerts and reminders should meet 14 CFR Part 23.1322, 25.1322, 27.1322 or 29.1322 as appropriate. Their intent should be generalized to the use of colors on displays and controls (AC 120-76A, 10.d (1))
- ❖ Red should be used only for warnings (AC 120-76A, 10.d (1))
- ❖ Amber should be used only for cautions (AC 120-76A, 10.d (1))
- ❖ Other colors should be sufficiently distinct from red/amber for use (AC 120-76A, 10.d (1))
- ❏ Alerts and reminders should be consistent with AC 25-11, 14 CFR Part 23.1311a, AMJ 25-11
- ❏ Alerts should be integrated or compatible with other flight deck alerts (AC 120-76A, 10.d (1))
- ❏ Messages should be prioritized and prioritization scheme should be documented and evaluated (AC 120-76A, 10.d (1) and AC 120-76A, 10.d (2))
- ❏ Strong attention-getting techniques (e.g., flashing or bright text) should be avoided (AC 120-76A, 10.d (1))
- ❏ During high workload phases of flight:
    (a) Required flight information should be continuously present and unobscured, except those that indicate failure or degradation of the EFB application (AC 120-76A, 10.d (1))
    (b) Messages should be inhibited, except those that indicate failure or degradation of the EFB application (AC 120-76A, 10.d (1))

2.4.14   Supplemental Audio

- ❏ Supplemental audio should be avoided in flight
- ❏ Users should be able to control the volume
- ❏ Users should be able to turn off the supplemental audio
- ❏ Objects with supplemental audio should be coded so the user knows of the associated audio before activating it
- ❏ Supplemental audio that is solely audio should have text description available
- ❏ Users should be able to stop the supplemental audio at any time

WORKLOAD

2.1.1   Workload

- ❏ Flight crew workload and head-down time should be minimized (AC 120-76A, Section 10.c)

3 Electronic Documents

Formatting/Layout

3.2.1 Consistency of Information Structure
- ❖ The information structure of the electronic document should be consistent with that of the hard copy

3.3.1 Visual Layout and Structure
- ❑ Windows and frames should be placed and used consistently
- ❑ Sections of text should be separated with plenty of white space
- ❑ Data should be formatted into short segments, where possible

3.3.2 Minimum Display Area and Resolution
- ❑ The minimum document display area and resolution should be specified by the manufacturer
- ❑ Operators should meet the manufacturer-specified display area and resolution requirements for training and operational use

3.3.3 Off-Screen Text
- ❖ The existence of off-screen content should be indicated clearly and consistently (AC 120-76A, 10.b (7))
- ❑ Whether it is acceptable for parts of the document to be off-screen should be based on the application and intended function (AC 120-76A, 10.b (7))
- ❑ Information regarding the document length and the current place within the document should be constantly available

3.3.4 Active Regions
- ❖ Active regions should be clearly indicated (AC 120-76A, 10.b (8))

3.3.6 Figures
- ❖ The electronic version of a figure should show all the content in the paper version
- ❖ The entire figure should be viewable at once, even if all the details are not readable
- ❖ All the details should be readable, although the entire figure may not be visible when doing so
- ❑ Figures should be displayed in their entirety with all details readable whenever possible
- ❑ Text information should be provided for each figure, independent of whether the figure is shown in full, or marked by a placeholder
- ❑ The user should be able to configure the figure for optimal viewing
- ❑ If zooming is supported, discrete zoom levels should be available (e.g. view whole page) and the current zoom level should be displayed at all times

Interactions: Accessing Functions and Options

3.4.1 Moving to Specific Locations
- ❑ The cursor should be visible at all times (AC 120-76A, 10.b (7))
- ❑ If links are supported:
    - Entries in the table of contents should be linked to its location in the text
    - Cross-references should be linked to each other within a document
- ❑ Users should be able to return to the previous location in one step

3.4.2 Managing Multiple Open Documents
- ❖ The active document should be indicated continuously (AC 120-76A, 10.b (9))
- ❖ The user should be able to choose the active open document
- ❑ A master list of all open documents should be available

3.4.3 Searching
- ❏ Search functionality should be available
- ❏ Users should be able to select the document(s) to include in the search

General Principles

3.5.1 Printing
- ❖ Pages or sections selected for printing should be clearly indicated
- ❖ The user should be able to terminate printing immediately
- ❏ Users should be able to select document subsets for printing
- ❏ The printed document should have the same visual structure as the EFB electronic document

3.5.2 Animation
- ❖ Start/stop functionality should be provided. The user should be able to stop the animation at any time
- ❖ Text describing the animation should be available even if the animation is not running
- ❏ Animation should not be overused
- ❏ If supplemental audio is provided, control of the audio and video should be integrated

4 Electronic Checklist Systems

Formatting/Layout

4.2.2 Information and Visual Layout/Structure of Electronic Checklists
- ❖ The resulting crew actions called for in the checklist should be identical for paper and electronic versions
- ❑ Layout of items should be similar to the paper version. Headings, sub-headings, and titles should be consistent (CAP 807)
- ❑ The format of the electronic checklist should make it clear which challenge is associated with which response (CAP 708)

4.3.2 Managing Checklists
- ❖ The checklist title should be displayed above the items and be distinguished throughout the checklist
- ❑ Parent-child checklists should be integrated into a single checklist
- ❑ If more than one checklist can be open at once, a master list of checklists should be available

4.3.3 Managing Non-Normal Checklists
- ❑ All checklists associated with on-going non-normal conditions that are sensed should be listed on one master list
- ❑ A master list should indicate the status of each checklist

4.3.6 Closing All Checklists
- ❑ The ECL should allow a state where no checklists are open
- ❑ The system should give a positive indication that no checklists are open; a blank screen is not sufficient

4.4.2 Displaying Item Status
- ❖ Item status, if available, should be clearly indicated.

4.4.4 Specifying Completion of Item
- ❑ The completion status of each checklist should be indicated clearly
  (see also Interactions: Accessing functions and options)

4.5.4 Checklist Branching
- ❑ The selected branch should be clearly indicated
  (see also Interactions: Accessing functions and options)

Interactions: Accessing functions and options

4.3.1 Accessing Checklists
- ❖ All supported checklists should be accessible for reference/review at any time while the system is active
- ❑ Normal checklists should be accessible in accordance with the normal sequence of use
- ❑ Electronic checklists should be as quick and accurate to access as paper checklists
- ❑ The ECL system should open checklists only upon crew request

4.3.2 Managing Checklists
- ❖ The title of each open checklist should be visible continuously
- ❖ If more than one checklist can be open at once, other checklists should be accessible without closing the displayed checklist
- ❖ If more than one checklist can be open, the user should be able to select which one is active
- ❖ If a checklist is a "child" of another checklist, the user should be able to select whether the parent or child is active
- ❑ A placeholder should be used to indicate which item was active prior to leaving the checklist
- ❑ The crew should be able to reset the checklist with a simple input

4.3.4 Lengthy Checklists
- ❖ The user should be able to look ahead (e.g., page down) without changing the active item
- ❏ Information regarding the length of the checklist, the user's current position within the checklist, and how much of the checklist has been completed should be continuously available
- ❏ It should not be possible to change the status of off-screen items
- ❏ If the active item is off-screen and the user makes an "item completed" entry, an error message should appear or the active item should be called into view

4.3.5 Closing or Completing a Checklist
- ❖ If item status is tracked and the user attempts to close an incomplete checklist, the system should provide an indication that the checklist is incomplete and present any deferred/incomplete items for review
- ❖ The user should be able to close incomplete checklists after acknowledging this indication
- ❏ If item status is tracked, a positive indication should be presented when the entire checklist, as well as each item, is completed
- ❏ The action for closing/completing a checklist should be distinct from the action for marking an item as complete

4.4.1 Indicating the Active Item
- ❏ The ECL should track and indicate the active checklist item
- ❏ When returning to an incomplete checklist, the item active prior to the move should again be active

4.4.3 Moving Between Items Within a Checklist
- ❖ The active-item pointer should be moved to the next item with a simple action
- ❖ Returning to a previous item should not change the status of any item
- ❏ If the status of individual items are tracked, the user should be able to:
  (a) Move from uncompleted items, changing their status to deferred
  (b) Move to the next item automatically after completing an item
- ❏ The user should be able to quickly select one item after another; system processing should not induce delays

4.4.4 Specifying Completion of Item
- ❖ User actions to mark an item as complete should be simple
- ❖ Completed items should not be removed from the screen immediately. The crew should be able to review the item and undo their action, if necessary
- ❏ If the system indicates active items:
  a) The next item in the list should become active when an item has been completed, unless it is on the next page. A separate action should be required to move to the next page
  b) Moving to the next item without completing the current item should require an input distinct from that of specifying the item as complete
- ❏ An *undo* function should be available
- ❏ The completion status of each checklist should be indicated clearly
  (see also Formatting/Layout)

4.5.1 Links Between Checklist Items and Related Information
- ❏ The navigation between links in the ECL and related information needs to be simple and clear
- ❏ Related information should appear in a single window or area of the screen. Hyperlinks from the related information should be shown in the same window or area

4.5.2 Links to Calculated Values
- ❑ If the EFB provides calculation worksheets and allows integration between the application hosting the ECL and the application hosting the calculation worksheets, then:
    i. Direct access to the appropriate worksheet should be provided for all items that can be calculated. This should be available for initial calculations and subsequent review/modifications
    ii. The user should be able to return easily to the checklist item from which the worksheet was accessed
- ❑ Calculated ECL values should appear in the corresponding checklist location. These fields should be blank prior to inserting the calculated value

4.5.4 Checklist Branching
- ❑ The user should be able to backup and select another decision branch
- ❑ Items not on the selected branch should not be selectable
- ❑ The selected branch should be clearly indicated
  (see also Formatting/Layout)

General Principles

4.2.1 Checklists Supported by the ECL System
- ❖ If normal checklists are supported, then *all* normal checklists should be supported
- ❖ If non-normal emergency checklists are supported, then *all* non-normal checklists should be supported
- ❖ Similar requirements apply for other checklist categories
- ❑ The ECL system should indicate the location of unsupported checklists in the paper document
- ❑ Non-normal checklists should retain as much commonality with normal checklists as possible

4.5.3 Task Reminders
- ❑ Reminders for high priority, time-critical tasks should be displayed constantly once in progress and should attract attention when delayed actions should be performed
- ❑ If multiple task reminders can be shown, crews should be able to determine how many are in progress and to what tasks they refer

5 Flight Performance Calculations
Interactions: Accessing Functions and Options

5.1.5   Modifying Performance Calculations
- ❑   The user should be able to modify previously computed results quickly
- ❑   Output relevant to earlier calculations should be erased once the user begins modifying those calculations

Error Handling and Prevention

5.1.2   Data-entry Screening and Error Messages
- ❑   The EFB should not accept user-entered data that is of incorrect format or type. Error messages should point out suspect entries and specify the expected data type. (AC 120-76A, Section 10.d (3))
- ❑   The system should detect input errors as early as possible during data entry (AC 120-76A, Section 10.d (3))
- ❑   The system should *only* discard erroneous input errors and not the whole set of entries related to the task in progress
- ❑   The system should present an error message when required values are missing; this error message should contain the name of the required value, using the label from the input field

General Principles

5.1.1   Default Values
- ❑   Blank data entry fields should be used to indicate that there is no system assigned default value

5.1.3   Support Information for Performance Data Entry
- ❖   The units of each variable should be clearly labeled
- ❑   Labels, formats, and units of variables should match that in other sources (e.g., paper reports, flight deck systems)
- ❑   Related information for cross-checking should be in view or easily accessible

6 Electronic Charts

Formatting/Layout

6.2.7 Orientation of Electronic Charts
- ❖ Orientation of the charts should be indicated continuously
- ❖ When charts are oriented with respect to directionality (e.g., track/heading), and directionality information becomes unusable, it should be clear to the pilot that that information is not available
- ❑ When charts are oriented with respect to directionality (e.g., track/heading), and directionality information becomes unusable,
  (a) The crew should be notified of the unusable directionality and informed that the charts must revert to north-up orientation.
  (b) After crew acknowledgement of the failure, the charts should revert to the north-up orientation, the chart orientation indicator should be updated, and any cues that could imply directionality should be removed
- ❑ Text and symbols other than those designed to reflect compass orientation should remain upright at all times
- ❑ Crew input should be required to change the orientation of the charts

Interactions: Accessing functions and options

6.2.5 Basic Zooming and Panning
- ❑ If zooming is supported, then panning should also be supported, and vice versa
- ❑ The chart's visual edges should be clearly marked. Visual edges should be shown only when no more information is outside that area
- ❑ When panning, the user should know which way to move to bring more of the chart into view
- ❑ Panning to an area where no portion of the chart will be displayed should be prevented
- ❑ If the user can change zoom levels, the user should be able to return to a default view easily
- ❑ If the display can be panned, the user should be able to return to a default view easily
- ❑ Zooming and panning should not result in lengthy processing delays

6.2.9 Access to Individual Charts
- ❖ The currently selected chart's label should be displayed continuously
- ❑ The system should allow rapid access to pre-selected charts
- ❑ The chart application should help the crew ensure that the correct chart was selected and allow corrections to be made quickly when an error occurs
- ❑ Multiple search methods should be supported
- ❑ Search results should be ordered with its best guesses at the top of the list and least likely to be used charts at the bottom
- ❑ Selection of alternate runways should be facilitated during approach

6.2.11 De-cluttering and Display Configuration
- ❑ The pilot should not be able to de-clutter safety critical display elements without knowing they are suppressed
- ❑ Changing map scale, orientation, and other options and settings should not induce significant levels of workload
- ❑ The information prioritization scheme should be documented

Error Handling and Prevention

6.2.2 Updates to Electronic Charts
- ❑ Corrections/updates should be made directly within the electronic chart application, unless they are temporary
- ❑ Corrections/updates that are of high priority or time-sensitive should not be made via paper notifications

6.2.4 Scale Information
- ❖ Scale information should always be visible for charts drawn to scale
- ❖ Scale information should be accurate. Scale information should be updated when the display is zoomed
- ❖ Static scale information should be removed unless it is always accurate
- ❖ Charts drawn "not to scale" should have a label indicating that fact continuously

6.2.10 Knowledge and Display of Own-Aircraft Position
- ❖ Display of ownship should not be supported on non-georeferenced or not-to-scale terminal charts
- ❖ See TSO C-165 and DO-257A for other applicable requirements
- ❑ The range of display zoom levels should be compatible with the position accuracy of the ownship symbol.
- ❑ An indication of ownship position should be provided if the chart is zoomed or panned such that ownship is not in the current view

General Principles

6.2.1 Transition from Paper to Electronic Charts
- ❑ Information structure of electronic charts should match that of paper charts
- ❑ Visual structure of electronic charts should be compatible with paper charts

6.2.3 Hard Copy Backups of Electronic Charts
- ❑ If the hard copy is used as a backup, it should be of sufficient quality to be used as effectively as the original paper chart. In particular:
    (a) The hard copy should be legible; all chart details should be visible
    (b) The quality of the paper should be acceptable for normal use
    (c) Color information should be distinguishable in the monochrome hard copy
    (d) All the chart information should fit on one printed page
    (e) The hard copy should be at least as large as a standard paper chart
    (f) The user should be able to select the size of the hard copy

Appendix B: EFB User-Interface Assessment Tool

EFB User-Interface Assessment Tool

HARDWARE CONSIDERATIONS

- Physical Ease of Use
  — Input devices and display, accessibility of controls
- Labels and Controls
- Lighting Issues (day vs. night use)
  — Brightness adjustment, illumination of labels
- Amount of feedback, potential for errors

SOFTWARE CONSIDERATIONS

Symbols and Graphical Icons
- Clarity of intended meaning, confusability
- Legibility and distinctiveness

Formatting/Layout
- Fonts (size, style, case, spacing)
- Arrangement of information on the display
  — Consistency with user expectations and internal logic

Electronic Documents
- Indication of active regions and off-screen material
- Figures/tables
- Page format
- Structure and organization, consistency with hard copy

Electronic Checklists
- Display of item status, e.g., open, deferred, completed
- Indication of checklist status, e.g., open, closed, completed, active
- Formatting (e.g., associating challenges with responses)
- Consistency with hard copy

Electronic Charts
- Formatting
- Structure and organization, consistency with hard copy

Interaction (Accessing functions and options)
- Home pages and ease of movement between pages
- Number of inputs to complete a task
- Ease of accessing functions and options
- Feedback (system state, alerts, modes, etc)
- Responsiveness
- Intuitive logic

Electronic Documents
- Moving within a document, moving between documents
- Identifying open documents, identifying current document
- Zooming
- Search functionality

Electronic Checklists
- Accessing checklists and moving between checklists
- Managing checklists, e.g., parent-child relationships, master list
- Identifying open checklists, identifying current checklist
- Moving between items
- Linking between items, calculated values, other related information

Flight Performance Calculations
- Modifying performance calculations

Electronic Charts
- Access to charts
- Identifying open charts, identifying current charts
- Zooming and panning
- De-cluttering and display configuration (e.g., scale, orientation)
- Search functionality

Error handling and prevention
- Susceptibility to error (mode errors, selection errors, data entry errors, reading errors, etc.)
- Correcting errors (e.g., cancel, clear, undo)
- Error messages

Electronic Charts
- Updating chart information
- Scale information

Flight Performance Calculations
- Data entry

Multiple Applications
- Consistency and compatibility across applications
- Identifying current position within system
- Ease of switching between applications

Automation (if any)
- Is there enough? Too much?
- Is it disruptive/supportive? Predictable? User control over automation? (e.g., manual override)

General
- Consistency of controls/elements; are they distinctive where appropriate?
- Visual, audio, and tactile characteristics
- Use of color (esp. red and amber) and color-coding
- Amount of feedback (system state, alerts, modes, etc)
- Clarity and consistency of language, terms, and abbreviations
- End-user customization (if any)

Electronic Documents
- Printing (if available), printouts
- Animation (if any)

Electronic Checklists
- Set of checklists that are supported
- Presentation of task reminders (if any)

Flight Performance Calculations
- Unit labels
- Default values

Electronic Charts
- If own-aircraft/ownship display, see TSO C-165
- Printing (if available), printouts

WORKLOAD
- Problem areas

OTHER

Appendix C: Guide for Developing Simulator and Validation Flight Scenarios

Simulator and/or in-flight validation tests may be needed to fully determine the suitability of an EFB (see AC 120-76A Paragraph 12 (j), pp. 21-22). The following event-based scenarios may be helpful in constructing EFB validation scenarios. The examples below are only generic suggestions; each operator's proposed EFB functionality and software will vary and scenarios should be customized for the particular situation by the inspector and applicant.

Where appropriate, some of the tests could be conducted as part of the operator's 6-month field test of the EFB. If the operator has approved line operational scenarios, the EFB could be integrated into these existing scenarios to provide a basis for evaluation. Some of the suggested simulated emergency procedures may only be appropriate in a simulator or training device. The most appropriate means for the validation should be determined together by the inspector and applicant.

At the end of the validation flight(s), it should be evident that, as applicable, information provided by the EFB is at least equal to that obtained from pre-EFB methods.

1. **Scenarios**

The validation flight scenarios should be used to ensure that EFB use has been adequately transitioned into the operator's overall training and operations programs. The scenarios should not be combined so as to overload an individual pilot or crew. Note that the tasks below do not specify how the EFB will be used in detail; they merely specify what the crew must accomplish. In some cases, the task will be completed entirely with an EFB, and in other cases, the EFB may be used together with other sources of information (e.g., paper charts or documents), depending on the capabilities of the EFB and its operational implementation.

Six classes of scenarios are presented below, based on the phase of flight.

   a. Preflight planning. Observe crew actions and EFB use in preparing for the flight (e.g., in calculating aircraft weight and balance, takeoff, climb and maneuvering speeds).

   - Compare values from the EFB with values computed from previously approved methods. Check at least three samples throughout the range of performance (i.e., minimum to maximum).

   - Observe how the pilot/crew maintains critical data for immediate reference (e.g., fuel quantity, "V speeds", etc.).

   - During taxi, introduce a runway change and, if an EFB is used for critical aircraft system information, initiate the need to reference one or more applicable items such as an airframe deicing fluid requirement, MEL item, etc.

   - Introduce time critical adjustments prior to block out/taxi and takeoff (e.g., fuel, passenger load, etc.).

   b. Takeoff. Observe crew actions and EFB use during several types of departures.

- Combine a complex Standard Instrument Departure (SID) or Departure Procedure (DP) with an abnormal or emergency event during the departure climb-out.
- Establish take-off on a runway that requires recognition/briefing special operator engine-out procedure (if applicable).
- Introduce an engine failure or other significant emergency that requires a return to the departure or alternate departure airport.
- On takeoff roll, observe actions taken when all EFB screens fail ("blank out") prior to V1 (or rotation, as applicable).
- Immediately after takoff, observe actions taken when all EFB screens fail ("blank out"), or when one of two EFBs fail, requiring one pilot to rely on the EFB of the other pilot.

c. **Level-off/Cruise.** Observe crew actions and EFB use during abnormal situations in cruise.

- Initiate an engine-failure/fire with possible condition of destination below weather minimums. (If applicable, require drift down solution.)
- Initiate electrical smoke in the cockpit requiring use of smoke mask/goggles while completing checklists, using EFB for approach briefing, etc.
- Initiate abnormal condition requiring EFB for reference of MEL or other procedural guidance (as applicable).
- If cabin crew interact with the flight crew through EFB in anyway, introduce an abnormal situation, medical emergency, maintenance item, etc. (These could be added to any other flight phase scenario, if applicable.)

d. **Descent.** Observe crew actions and EFB use during preparation for landing.

- During approach to landing, introduce a runway change, holding, and/or the need to re-compute landing weight and V speeds.
- During descent, tell the crew that reported runway conditions require reference to operational limitations due to contamination, wind, etc.

e. **Approach/Landing** Observe crew actions and EFB use under poor weather conditions, or to airports with complex taxi routes.

- During approach/landing, tell the crew that conditions require reference to SMGCS taxi routing or a complex clearance.
- Initiate an ATC request for specific taxiway turn off during rollout after landing.

f. **Destination Ground Operations:** Observe crew actions and EFB use during ground operations.

- Initiate EFB partial failure or simulate possible erroneous output requiring maintenance discrepancy to be entered.

2. Expanded Sample Scenarios

The EFB validation-flight scenarios given above could be affected by different factors, such as:

- Software: Type of EFB software application(s) (Type A, B, or C)
- Hardware: Class of EFB hardware (Class 1, 2, or 3), which includes factors such as location in the flight deck, and connectivity to other aircraft systems.
- Aircraft/Operations: Type of aircraft and operations (e.g., single pilot vs. dual pilot, single EFB vs. dual EFB)
- Weather: Weather conditions (e.g., visual vs. instrument, or very low visibility)

The four examples below illustrate how these factors could affect the use of the EFB in more detail. In each example, various conditions are assumed, and consequences for the EFB evaluation are explored.

   a. **Preflight Planning.** Observe how pilot/crew maintains *V speeds* for immediate reference. In particular, V speeds must be visible and directly in front of the crews during takeoff (regardless of the type of operation).

   - Software: Assume Flight Performance calculations, a Type B application
   - Hardware: Class 1 and 2 EFBs are generally not located directly in front of the pilot during takeoff. Therefore, V speed calculations completed on Class 1 or 2 EFBs would need to be transferred from the EFB (e.g., onto a display bug, or piece of paper) and placed in the pilot's forward field of view for takeoff. A Class 3 EFB may have communication capabilities so that V speeds calculated on the EFB could be transferred electronically to displays that are directly in front of the crew.
   - Aircraft/Operations: This task applies to all operations.
   - Weather: Performance of this task would not vary with weather.

   b. **Takeoff.** Assume that the EFB is displaying an electronic chart during takeoff. The EFB goes blank prior to V1 (or rotation, as applicable).

   - Software: Assume Type B (Interactive) Electronic charts application
   - Hardware: A Class 1 EFB cannot be in use during takeoff, and so this example applies only to Class 2 and 3 EFBs.
   - Aircraft/Operations: This task is applicable to all aircraft/operation during takeoff.
   - Weather: In visual flight conditions, the pilot could continue the takeoff without the information provided by the EFB. In low visibility or instrument conditions, considerations should be given to returning to the field or diverting to an alternate airport.

   c. **Level-off/Cruise.** Initiate a diversion to a destination that is below weather minimums. The diversion could be caused by weather, a maintenance issue, or an emergency, such as an engine-failure/fire.

   - Software: Could have electronic checklists, electronic charts, electronic documents, or any combination of these on the EFB. The electronic checklists may or may not include emergency checklists. The applications could share information between them, or be completely independent from one another.

- Hardware: EFB could be of any hardware class. Single or dual EFBs could be present. If there are dual EFBs, they could be independent so that the pilot-flying and the pilot-non-flying could refer to different information.

- Aircraft/Operations: In a single-pilot, single-EFB condition, it would be difficult to use an EFB effectively to manage an emergency situation. In a dual-crew, dual-EFB, Class 3 system with fully integrated electronic emergency checklists, the EFB could make an emergency situation easier to handle.

- Weather: During turbulence, managing the EFB could be more difficult. Depending on the weather, alternate approach procedures may need to be considered, implying heavy use of an electronic chart application.

d. **Descent/Approach/Landing** During descent into an airport experiencing low visibility conditions, the pilot/crew needs to access information about operational limitations. During approach/landing into the field, conditions require reference to SMGCS (low visibility) taxi routing or a complex clearance.

- Software Application(s): Assume Type B (Interactive) Electronic charts application. Relevant documents could also be available on the EFB.

- Hardware: EFB could be of any hardware class. Although Class 1 hardware is generally not permitted to be used at low altitudes, it could be used during the beginning of the descent, and during surface operations.

- Aircraft/Operations: This scenario is applicable for evaluating EFB use by airlines landing at airports with SMGCS routes. (CAT II and III conditions require special ground routes, equipment, and charts.) The SMGCS procedure could be displayed on an EFB in an electronic chart application. Because these charts show complex taxi routes, the crew may need to zoom in and out of the chart often to maintain a view of the route, implying increased workload (in an already difficult situation). The SMGCS procedures may also need to be in the pilot's primary field of view. This could be a difficult scenario for a single pilot who is using a Class 1 EFB.

- Weather: Reported runway conditions could require reference to documents to obtain information about operational limitations due to contamination, wind, etc. during descent. Turbulence during the descent/approach could also affect use of the EFB.

Appendix D: Operational Evaluation Questions

This appendix contains a comprehensive list of questions for consideration during a "desk-top" EFB evaluation (i.e., an evaluation conducted outside the context of a simulated or actual flight). The questions are designed to promote a thoughtful structured exploration and review of the EFB system from a human factors perspective. In cases where the FAA team finds that a system shows weaknesses or limitations, or where the FAA team simply cannot predict how well the system will perform, mitigations should be developed in consultation with the applicant.

These questions are intended to address a wide variety of operators/equipment. The FAA inspection team should customize its use of these questions. For example, for simple EFBs (e.g., Class I, Type A), certain questions may not be applicable in the view of the FAA inspection team. Some questions have sub-items, which could be questions or considerations that clarify and expand upon the primary question, but some sub-items may not be applicable to the specific situation.

The appendix is divided into three subsections. The first, Section A, covers general operational evaluation questions. This section is for use by *both* the Aircraft Evaluation Group (AEG)/Aviation Safety Inspector (ASI), and the Flight Standards District Office (FSDO)/Principal Inspector (PI). Within Section A, there are five main sections:

1. General EFB System
2. Electronic Documents
3. Electronic Checklist Systems (ECL)
4. Flight Performance Calculations
5. Electronic Charts

Of these five main sections, the first (General EFB System) is the largest. Within this large section, topics are further subdivided into the following sections: General Considerations, Physical Placement, Training/Procedures Considerations,

Software Considerations, and Hardware Considerations.

The second part of this appendix, Section B, includes additional questions that are appropriate during an evaluation by the AEG/ASI. In general, questions that are specific to the AEG/ASI are related to initial installations and training for a given aircraft. Some of the AEG/ASI questions provide for a more thorough evaluation, appropriate for EFBs that will be used in a more complex manner. For example, this section contains detailed questions on applications such as Electronic Charts, Flight Performance Calculations, and Electronic Checklists. Section B is not intended for use by the FSDO/PI.

The last part of this appendix, Section C, contains additional questions that are appropriate during an evaluation by the FSDO/PI. Questions that are specific to the FSDO/PI are generally related to documentation and long-term use of the EFB (e.g., during the 6-month operational evaluation). Questions in Section C are not appropriate for the AEG/ASI.

A. General Operational Evaluation Questions

This section covers general operational evaluation questions for an EFB system. This section is for use by both the Aircraft Evaluation Group (AEG)/Aviation Safety Inspector (ASI), and the Flight Standards District Office (FSDO)/Principal Inspector (PI). This section is divided into the following subsections: General Considerations, Physical Placement, Training/Procedures Considerations,

Software Considerations, and Hardware Considerations.

1. GENERAL EFB SYSTEM

1.1 General Considerations

1.1.1 Workload

See also 1.1.1 in Section B (p. 15).

a) How does the workload required for completing a task with the EFB compare with the workload for completing the task with a conventional method?
   — If there is an increase in the workload of completing a task with the EFB relative to alternative methods, is this increase acceptable?

b) Are additional policies or procedures required to safely accommodate the EFB?
   — What are they?
   — Are they adequate?

c) Is there any impact to crew workload from an EFB failure?
   — If yes, is the impact acceptable?

d) Are there any aircraft system failure procedures (i.e. electrical smoke, fire, etc.) that could render the EFB unusable?
   — If yes, is this incorporated into procedures, checklists, etc.?

1.1.2 Using EFBs During High Workload Phases of Flight

a) Does the use of the EFB impose additional workload during a high workload phase of flight?
   — For example, are complex, multi-step data entry tasks avoided during takeoff, landing, and other high workload phases of flight?
   — Do company procedures mitigate workload issues?

b) If the EFB is designed for use during high workload phases of flight (including takeoff and landing), is it secured within the aircraft?

c) Are additional policies or procedures required to safely accommodate the EFB in high workload phases of flight (e.g., must approach briefings be accomplished earlier en route, restrictions placed on multi-function use, etc.)?
   — What are they?
   — Are they adequate?
   — Are they included in pilot/crewmember training?

d) Are there procedures, policies, or built-in limits on use of the EFB to ensure that pilots do not become distracted during high workload phases of flight?

1.1.3 Keeping EFB Content/Databases Current and Ensuring Integrity of EFB Data

a) *For each of the applications* on the EFB, what are the *procedures* for keeping the databases/stored data accurate, current, complete, and uncorrupted?
   — Who modifies the content/databases and how?
   — How are changes to content/databases documented?
   — How are crews notified of updates?

- If any applications use information that is specific to the airplane type or tail number, are there procedures to ensure that the correct information is installed on each airplane?
- Are operational control procedures consistent with regulations concerning preventative maintenance?

b) What procedures are in place to avoid corruption/errors during changes to the EFB system?

c) If there are multiple EFBs on the flight deck, are their procedures to ensure that they all have the same content/databases installed?

1.1.4 Compatibility and Consistency with Flight Deck Systems and Other Flight Information

a) Are there any noticeable conflicts between the EFB and flight deck interfaces, or is the user interface of the EFB generally compatible with the flight deck? (In order to be "compatible," the EFB user interface should not be in direct conflict with other systems.)

- If there are conflicts between the EFB and flight deck interfaces, how significant are they? Is the user EFB interface still acceptable?

b) Does the EFB minimize the potential for crew error by using terms, icons, color codes, and symbols that are consistent with flight deck systems and other sources of flight information? Note that, in order to be "consistent", the EFB user interface should match the other systems.

1.1.5 Use of the EFB with Other Flight Deck Systems

a) Are there procedures to ensure that the crew knows what flight deck system information is to be used if there is any redundancy with the information from *any application* on the EFB?

- For example, if the EFB computes data that the FMS also computes, which is primary?
- What are the procedures for establishing which source of information is primary?

b) What procedures does the crew follow if there is a disagreement between the EFB and other flight deck systems, or between multiple EFBs?

c) Is a backup source of information necessary?

- Under what conditions will the backup source of information be used?
- What are the consequences of using backup information?

1.1.6 Lighting Issues

a) Can the EFB screen be read under a variety of typical flight-deck lighting conditions?

- If no, what mitigations are available for making it possible to read the EFB screen? Are these mitigations acceptable?

b) If the EFB is to be used outside the flight deck, can the EFB screen be read under outdoor lighting conditions?

c) Can the user adjust the screen brightness and contrast?

- Does the EFB adjust screen brightness automatically, and if so, is the adjustment acceptable?

d) Are buttons and labels adequately illuminated for all environmental conditions (e.g., day, night, weather)?

e) If predetermined settings for illumination are required, are they incorporated in pilot procedures, or checklists?

1.1.7 System Shutdown

a) Are unique procedures for shutting down the EFB necessary (e.g., over and above normal aircraft parking/shutdown)?

- What are they?
- Are they designed for long-term stability of the EFB and ease of crew operation?
- What happens if the crew cuts power to the EFB instead of shutting it down properly?
- Are previous users' data entries cleared upon shutdown so that the system starts up in a predictable state?

b) Does the EFB function correctly when rebooted?

1.1.8 Failures

a) What are the failure modes for the hardware and software?
- How does each type of failure affect crew and/or aircraft operations?
- Should there be any MMEL/MEL items to handle these failures?

b) Are failures obvious to the crew?
- Is the nature of the failure clear?

c) Are failures handled with minimum impact to crew tasks and workload?
- Are there special EFB checklist failure items that must be incorporated into FAA approved checklists?

d) Are there procedures in place for the crew in case a failure occurs?
- If the EFB "hangs", fails to respond to crew input, or displays error or fault messages, are the means of recovery easy to remember and apply?
- Does the crew have to remember any arbitrary procedures or refer to paper documentation in order to restart the EFB?

1.2 Physical Placement

1.2.1 Stowage Area

a) Is there a stowage area for the EFB? When the EFB is not stowed, is the securing mechanism in the stowage area unobtrusive?

b) When the device is stowed, does the combination of it and the securing mechanism intrude into any other flight deck spaces, causing either visual or physical obstruction of important flight controls/displays and/or egress routes?

c) Is the design of the stowage area acceptable?
- Does movement of the EFB to and from a stowage area require substantial effort, or substantially limit access to flight displays and controls?
- Is the securing mechanism simple to operate for a wide population of users?
- Are the device and/or the stowage area easily damaged under normal usage?

1.2.2 Use of Unsecured EFBs (includes Operations Procedures under MEL)
a) Does the pilot have adequate access to flight controls and displays when the unsecured EFB is in use?
b) Is there an acceptable place to put an unsecured EFB when in use?
c) Is there an acceptable place to put an unsecured EFB when not in use?

1.2.3 Kneeboard EFBs
Note: The AEG would only evaluate kneeboard EFBs if a Type B application is supported.
a) Can the kneeboard EFB be positioned such that the pilot has full control authority?
b) Is the kneeboard EFB comfortable for the pilot to wear under normal conditions?
c) Are there special procedures in place for removal of the EFB during emergency landing or egress?

1.2.4 Design and Placement of Structural Cradle
See 1.2.4 in Section B (p. 15).

1.3 Training/Procedures Considerations

1.3.1 Training on Using EFB Applications
Is there a training program on how to display and interact with each of the individual applications (e.g., electronic documents, electronic charts, or electronic checklists)? Is it adequate?

— Do crews understand how to use any new or unique features of the electronic applications (e.g., do crews know how to use electronic document functions that do not exist for paper documents, such as hyperlinks and search)? Note: For Part 91 operators, refer to FAA Industry Training Standards (FITS) program.

1.3.2 Operations EFB Documentation and Policy
See also 1.3.2 in Section C (p. 1).
a) Is the documentation provided by the manufacturer with the EFB sufficient?
b) Are adequate MMEL/MEL items for the EFB in the manual?

1.3.3 EFB Training
See 1.3.3 in Section B (p. 15), and 1.3.3 in Section C (p. 1).

1.3.4 Fidelity of EFB Training Device
Is the actual EFB used during training? If not, does the substitute EFB (training device) provide an adequate degree of fidelity?

— Does the training device simulate the key aspects of the task?

1.3.5 User Feedback
See 1.3.5 in Section C (p. 1).

1.4 Software Considerations

1.4.1 User Interface—General Design

See also 1.4.1 in Section B (p. 15).

a) Is the organization of the software adequate?

— For example, are the user interface, functions, function labels, and functional and navigation logic consistent with established user interface conventions for similar systems?

— Is any information expected by the crews missing or in a different place?

b) Was the layout of information on the screens adequate?

— For example, are similar or related fields, indicators, or controls located near each other? Are controls separated adequately if using the wrong one unintentionally has significant consequences?

c) Are common actions and time-critical functions easy to access?

1.4.2 General Use of Colors

See also 1.4.2 in Section B (p. 15)

a) Are red and amber/yellow used? If so, are they used appropriately? Red should be used only for warnings and amber/yellow only for cautions.

b) If multiple colors are used, can they all be seen and distinguished under the various lighting conditions in which the EFB will be used?

c) If colors can be customized, are there procedures or built-in limits that prevent defining color schemes that conflict with flight deck color conventions?

1.4.3 Symbols and Icons

a) Are symbols (e.g., graphical objects on an electronic chart) and icons (graphical controls) clearly depicted on the screen in all viewing conditions? That is, are the symbols and icons legible?

— Are their functions obvious?

— Are the symbols and icons distinguishable from one another?

b) Are any icons confusing? Is training necessary to ensure that the icons are understood? (Icons are software-implemented controls that are represented on the screen by graphical pictures of limited size and resolution.)

— Does the initial EFB training adequately address icon meanings?

— Does the system provide information that explains each icon's meaning (e.g., a text label)?

c) Are the EFB icons and symbols compatible with those depicted on paper equivalents?

1.4.4 Legibility of Text—Characters, Typeface, Size, Width, and Spacing

Is the text easily readable?

— Do the characters stand out against the screen background?

— Are upper case and italic text used infrequently?

— Are the characters sufficiently large for normal viewing conditions?

— Is information that will be used in low-visibility conditions (e.g., emergency checklists) presented in text that is especially large and easy to read?

— If the text is too small to be read easily, it is it easy to zoom in on it to make it legible?

- Is the spacing between characters appropriate?
- Is the vertical spacing between lines appropriate?

1.4.5 Multi-Tasking
a) Is it easy to tell which application is currently open?
b) Can the pilot switch between applications easily?
c) Is an extra acknowledgement required to open applications that are not flight related?
d) Do all applications that are open at the same time function as intended on an individual basis?

1.4.6 Responsiveness
a) Does the system respond immediately to user inputs, e.g., by providing feedback?
- If processing is delayed, are busy indicators and/or progress indicators displayed?
- Are the indicators clear and useful to the pilot?

b) Does the system processing ever slow down to the point where normal use is impaired?

1.4.7 Alerts and Reminders
See 1.4.7 in Section B (p. 15).

1.4.8 Display of System Status
See 1.4.8 in Section B (p. 15).

1.4.9 Supplemental Audio and Video
Does the EFB support audio and/or video that are not associated with alerts, cautions or other critical system information? If yes,
- Does the operator have a policy regarding the use of this "supplemental" audio and/or video in flight?
- Does the user have control over when, and whether, the audio and/or video is activated?
- Is the audio audible in flight?
- Does the audio interfere with higher priority aural tasks (e.g., communications)?

1.4.10 Crew Confirmation of EFB Software/Database Approval
Is there a procedure for ensuring that data in use is approved for use in flight?
- Is the procedure for checking the EFB data approval consistent with standard operating procedures?
- Can the crew request revision information from the EFB? Is the revision information presented clearly?
- Are procedures in place so pilots know what to do if the database is not approved for use in flight?

1.4.11 Links to Related Material
Is access to related information supported?
- Are similar types of information accessed in the same way?
- Is it easy to return to the place where the user started from?

1.4.12 User-Interface Customization

a) If the crew (i.e., end-user) can customize the appearance of the EFB (not related to panning/zooming), is it easy to reset all parameters to their default values? [Note: Crewmember customization capability is not a recommended practice. Customization may have an adverse affect on items in Section 1.1, General Considerations]

— Is there a procedure or checklist item to ensure that crews clear all customized values?

— Does the EFB auto-reset to default values upon shutdown so that the system starts up in a predictable state?

— Does any customization have an adverse affect on items in Section 1.1, General Considerations?

b) Is the operator capable of customizing the appearance of the EFB?

— If yes, is the customization controlled through an administrative process?

— Does any customization have an adverse affect on items in Section 1.1, General Considerations?

1.5 Hardware Considerations

1.5.1 Display

See also 1.5.1 in Section C (p. 2)

Is the display acceptable for use of the intended applications? Consider its resolution, brightness, off-axis readability, etc.

— If artifacts appear on the display (e.g., ghost images or lines, jagged lines, or fuzzy images), do they impair the readability or functionality of the system?

1.5.2 Hardware Controls and Keyboards

a) Are controls labeled consistently and briefly for their intended function?

b) Can the user easily enter the most common types of input in any operational environment?

— Can crews use pointing and cursor control devices (if any) quickly, accurately, reliably, and repeatedly under all environmental and lighting conditions (e.g., turbulence, darkness)?

c) Is a keyboard appropriate for the task?

— Do the keys provide sufficient tactile feedback in all environmental conditions (e.g., turbulence)?

— Is key action firm enough to resist unintended actuation?

d) Is inadvertent activation of controls deterred?

— For example, do the physical keys provide tactile feedback?

— If a key is held down for a long time, is the input processed correctly? (For example, multiple entries may need to be discarded.)

1.5.3 Accessibility of Hardware Components

See also 1.5.3 in Section B (p. 16)

a) Are hardware components that are routinely used by the crew easy to access?
   — If not, is there any impact on flight task performance or safety?
b) Are the hardware components usable in the flight environment?
   — For example, will connectors stay in place after lengthy use in a vibrating environment or will a stylus remain functional?
   — If not, what mitigations are in place to ensure that the hardware components can be used?

2. ELECTRONIC DOCUMENTS

2.1.1 Training on Electronic Documents

Is there a training program on how to display and interact with electronic documents? Is it adequate?

2.1.2 Document Organization and Appearance

a) Can the crews find the material they are looking for?
   — Is the information organized in a way that makes sense to the crews?
   — Is the information arranged in a consistent way on the screen so that the crews know where to look for specific types of information?
   — Is it obvious when text is out of view? Is it easy to bring that text into view?
   — Can the crew tell where they are in relation to the full document?
   — Can the crew tell where they are in relation to the section of the document they are currently viewing?
b) Is the text of the document easy to read on the screen?
   — Is white space used to separate short main sections of text?
   — Is high priority information especially easy to read?
c) Are tables readable and usable?
   — How are especially long and complex tables handled?
d) Are figures readable and usable?
   — Can the entire figure be viewed at one time?
   — Can the crew zoom in to read details on the figure?

2.1.3 Interacting with Documents

a) Is it easy to move quickly to specific locations (e.g., to the beginning of a section, or to recently visited locations)?
   — Are active regions (e.g., hyperlinks) clearly indicated?
b) Is it easy to move between documents quickly?
   — Is it easy to tell what document is currently in view?
   — Is there a list of available documents to choose from?
c) Can crews search the document electronically?
   — Is the search technique adequate?

d) If animation is supported, does the crew have adequate control over it?
- Can the crew start and stop the animation as needed?
- Is there a text description of the animation that describes its contents (so the crews know its contents without running the segment)?

e) Is printing supported? If so, is it adequate?
- Can crews select a portion of a document to be printed?
- Is the hard copy usable?
- Can the crew terminate a print job immediately, if necessary?

3. ELECTRONIC CHECKLIST SYSTEMS (ECL)

An ECL is Type B software if the checklist is "interactive" (e.g., item status is tracked). Such systems need only AEG review for initial approval. The FSDO/PI may need to evaluate use of ECL during 6-month operational evaluation. For ECL that are essentially static images of paper checklists, the FSDO may need to review a subset of the questions below.

3.1.1 Training for Electronic Checklist Systems

Is there a training program on how to display and interact with electronic checklists? Is it adequate?

- Does using the electronic checklist produce the same crew actions that using the paper equivalent would?
- Are crews trained on how to use any new or unique features of the electronic checklists (i.e., functions that are not supported with paper checklists)?
- Are crews trained to know which checklists are supported electronically and which are not?
- Are crews trained to be aware of the limits of the ECL automation? In particular, are they trained on the limits of any ECL "sensing" functions?
- If the ECL senses aircraft status and uses this information to customize the checklists (e.g., by automatically selecting a decision branch), are any special training or procedures needed?

3.1.2 Access to Checklists

a) Is it easy to find and access specific checklists?
- Are normal checklists available in the appropriate order of use?
- Can checklists be accessed individually for review or reference?
- During non-normal conditions, are relevant checklists especially easy to access?

b) Is it easy to know where any given checklist will be found (on the EFB or on paper)?
- If the electronic checklist refers the crew to a paper document, is the location of that document provided within the electronic checklist?

3.1.3 Checklist Appearance

Is the layout and formatting of the ECL clear?
- Is the layout and formatting of the challenges and responses consistent with the paper checklist equivalent?

3.1.4 Managing Checklists

Can crews easily manage the checklists?

- Does each checklist have a constantly visible title that is distinct from other checklists?
- Can the crew easily pick which checklist they want to work on from a set of open checklists?
- Can crews page ahead to view items in a long checklist without changing the item they are actively working on?
- Can the crew close an incomplete checklist after acknowledging that it is not complete?
- Is it clear when no checklists are open?
- During non-normal conditions, does the system indicate which checklists need to be performed or possibly ignored?
- Does the ECL discourage two checklists (or more) from being in progress simultaneously?

3.1.5 Interacting with Checklist Items

See 3.1.5 in Section B (p. 16).

3.1.6 Interacting with Checklists

See 3.1.6 in Section B (p. 17).

3.1.7 Links Between Checklist Items and Related Information

See 3.1.7 in Section B (p. 17).

4. FLIGHT PERFORMANCE CALCULATIONS

Flight performance calculations are Type B software. Only AEG review is required for initial approval, although the FSDO/PI may need to observe use of this software during 6-month operational evaluation period. See Section B, p. 17 for suggested evaluation questions.

5. ELECTRONIC CHARTS

Electronic charts are Type B software if the pilot can pan and zoom to configure the view of the chart. Only AEG review is required for initial approval. The FSDO/PI may need to observe use of Electronic Charts during the 6-month operational evaluation period.

5.1.1 Training, Policy, and Procedures for Use of Electronic Charts

See also 5.1.1 in Section B, p. 18.

Is training required on the electronic chart application?

- Is the training adequate?
- Are crews trained on any new or unique features of the electronic chart function (i.e., functions that are not supported with paper charts)?
- Are crews aware of any differences in map scale, orientation, and database quality between the electronic charts and other similar flight deck displays (e.g., moving map displays, weather displays, or traffic displays)?
- If own-aircraft position is displayed, are pilots aware of the limitations of the display of own aircraft position?
- Are crews trained on operator policies pertaining to use of the electronic charts application?

5.1.2 Access to Charts

See 5.1.2 in Section B (p. 18).

5.1.3 Chart Appearance

See 5.1.3 in Section B (p. 19).

5.1.4 Interacting with Charts

See 5.1.4 in Section B (p. 19).

B. Additional AEG/ASI Operational Evaluation Questions

This section contains additional questions that may be appropriate specifically for evaluation by the AEG/ASI. In general, questions that are specific to the AEG/ASI are related to initial installations and training for a given aircraft. Some of the AEG/ASI questions provide for a more thorough evaluation, appropriate for EFBs that will be used in a more complex manner. For example, this section contains detailed questions on applications such as Electronic Charts, Flight Performance Calculations, and Electronic Checklists. References to other sections of this appendix are provided when particular topics are also covered elsewhere.

1. GENERAL EFB SYSTEM

1.1 General Considerations

1.1.1 Workload

See also 1.1.1 in Section A (p. 4).

Is an in-flight evaluation necessary? (An in-flight evaluation may be necessary if you are not able to adequately evaluate each function intended for this specific operation while on the ground.)

— If so, did the in-flight evaluation confirm that the overall workload is acceptable?

1.2 Physical Placement

1.2.4 Design and Placement of Structural Cradle

a) Does the structural cradle obstruct visual or physical access to flight controls and/or displays?

— Which controls/displays are affected, and how important are they during the different phases of flight in which the EFB will be used?

b) Does the structural cradle obstruct the emergency egress path?
c) Are there adjustment and locking capabilities for optimal viewing or storage?

— Are crews able to adjust and lock the EFB or their seat position for optimal viewing or for storage?

— Does the position for optimal EFB viewing/storage also provide comfortable and reasonable access to all flight controls during both on ground and in-flight operations?

d) Is there adequate room to manipulate the device controls and view its display?
e) Is the installation design acceptable for use in high workload flight phases?

— Consider ease of access if used during high workload flight phases.

1.3 Training/Procedures Considerations

1.3.3 EFB Training

See also 1.3.3 in Section C (p. 1).

What are the minimum training, checking and currency requirements?

— Is EFB training customized for new users?

1.4 Software Considerations

1.4.1 User Interface—General Design

See also 1.4.1 in Section A (p. 8).

Is the user interface internally consistent?

— Are there standard ways to perform common actions?

- Are a common set of controls and graphical elements used?
- Was a style guide followed when developing the user interface?

1.4.2 General Use of Colors

See also 1.4.2 in Section A (p. 8).

Are colors that convey meaning used in combination with other cues, such as shape?

- For example, could the pilot understand all the information even if the screen was black and white?

1.4.7 Alerts and Reminders

a) For installed systems, do EFB alerts and reminders meet the requirements in the appropriate regulations (specifically §§ 23.1322 or 25.1322, as noted in FAA AC 120-76A, Par 10)?

b) Is there an overall scheme for generating alerts/reminders (e.g., when will they appear, how are they prioritized)?

- Is it adequate/appropriate?

c) Are distracting flashing symbols avoided?

d) Are EFB messages inhibited during high workload phases of flight unless they pertain to the failure or degradation of the current EFB application?

1.4.8 Display of System Status

a) Are partial or full failures of the EFB clearly indicated with a positive indication, not lack of an indication?

b) Is the immediacy of the failure annunciation appropriate to the function that is lost or disabled? (For example, failures of low-criticality functions should not produce intrusive alerts.)

1.5 Hardware Considerations

1.5.3 Accessibility of Hardware Components

See also 1.5.3 in Section A (p. 11).

Are the connectors easy to use?

- Consider how long it takes to make the connections, how likely errors will be, and whether any special tools are required.

2. ELECTRONIC DOCUMENTS

No additional questions for an AEG/ASI review.

3. ELECTRONIC CHECKLIST SYSTEMS (ECL)

3.1.5 Interacting with Checklist Items

a) Is progress through the checklist clear to the flight crew?

- Is the active item clearly indicated?

b) Is item status tracked by the system and displayed to the crew (e.g., completed, deferred, or open)?

- Is item status displayed clearly under all lighting conditions?

c) Can the crew easily change the status of an item?

- Can the crew easily mark an item complete?
- After completing an item, does the next item in the list automatically become active?

- Can the crew defer the current item without completing it?
- Can the crew easily reset an item's status to "incomplete"?
- Can crews easily reset all items within checklist to "incomplete" in order to begin the checklist again?
- Is it possible to change an item that is not currently in view? If so, is the item that was changed brought to the crew's attention?

d) Can the crew easily move between items within a checklist?
- Can the crew easily move the active-item pointer to the next checklist item?
- Can the crew move backward to a previous checklist item without affecting the status of any item? If the user moves forward in the checklist, are deferred items marked appropriately?
- Does the active item change to the next one in the list after an item is completed? Is there a tendency to skip items when attempting to move to the next item?
- Is a separate action required to move to the next page after all the items on the current page are completed or deferred?

3.1.6 Interacting with Checklists

a) If the crew attempts to close an incomplete checklist, are they reminded to review deferred and incomplete items?
b) When finishing a checklist, is there a clear indication to the crew that all individual items in the checklist are complete, as well as an indication that the checklist as a whole is complete?
c) Does the checklist provide reminders for tasks that require a delayed action (e.g., dumping fuel)?
- Do the reminders clearly specify what to do?
d) Does the checklist visually highlight decision branches?
- If so, are the decision branches clear?
- Can the crew easily back up if they choose the wrong branch?

3.1.7 Links Between Checklist Items and Related Information

If the ECL provide links to useful, related information (e.g., links to worksheets or definitions):
- Is it easy to select what information to view?
- Can the user return to the checklist from related information in one step?
- Is the related information always shown in one window or area of the screen regardless of how many links were selected?

4. FLIGHT PERFORMANCE CALCULATIONS

4.1.1 Training for Flight Performance Calculations

Is there a training program on using the flight performance application? Is it adequate?
- Do crews know when they can (or should) use the flight performance application?
- Are crews aware of any assumptions on which the calculations are based? For example, are crews trained to identify and review default values and assumptions about the aircraft status or environmental conditions?
- Do crews know how to enter information required by the software (e.g., corrections for temperature, pressure altitude, braking action, etc.)?

— Do crews understand how to interpret and use results of the flight performance calculations? For example, will the results be entered into a flight management system?

— Are the roles of dispatchers and flight crews coordinated?

4.1.2 Data Entry

a) Does the system identify entries that are clearly of the incorrect format or type and generate an appropriate error message?

— Does the error message clarify the type and range of data expected?

— Are errors in data entry identified at the earlier possible point?

b) Are units for performance data clearly labeled?

— Do the labels used in the EFB match the language of other operator documents?

c) Is all the information necessary for a given task presented together, or easily accessible?

d) Are any data (especially defaults values) obtained from other flight deck systems?

— If yes, what is the backup plan for assigning these values if communication with the other system is lost?

4.1.3 Modifying Performance Calculations

a) Can the crews modify performance calculations easily?

— Is it especially easy to make changes that might be done at the last minute?

b) Are outdated results of performance calculations deleted when modifications are entered?

4.1.4 Aircraft Performance Documentation

What is the procedure for ensuring that, if necessary, EFB data can be stored outside of the device? (see 14 CFR Part 121.697)

5. ELECTRONIC CHARTS

5.1.1 Training, Policy, and Procedures for Use of Electronic Charts

See also 5.1.1 in Section A (p. 13).

For Part 121/135 operators, does the EFB policy specifically address the electronic charts application?

— Does the policy specify what other EFB functions or applications (if any) can be used while a procedure using the electronic charts is actively being flown?

— Does the policy address special procedures that may apply if the electronic chart application senses and uses aircraft state (e.g., ownship position) to customize its functions?

5.1.2 Access to Charts

a) Can crews find and display the charts that they are looking for quickly and accurately?

— Is there a way to pre-select specific charts for especially easy access during a particular flight?

— Can crews easily identify errors in chart selection?

— Is there more than one way to search for a chart?

— If a last minute change is necessary, can the crew easily handle a clearance/runway change?

b) If the chart application uses aircraft state (e.g., ownship position) to facilitate access to charts, does this function work adequately?
- Are appropriate charts brought to the crew's attention?
- Can the crew disregard and override system suggestions easily?

c) Are there procedures to ensure that all necessary navigation/approach charts appropriate for the flight are installed and available?

a)

5.1.3 Chart Appearance

a) Do the aeronautical charts conform to the guidelines of AC 211-2 "Recommended Standards for IFR Aeronautical Charts"?

b) Is chart scale information accurate and always visible?
- Is the scale indicator updated when the display is zoomed?
- Does the scale indicator stay in view as the display is panned?
- Is the potentially inaccurate static scale information (which comes as part of the chart database) removed from the display?

c) If electronic chart symbols are color-coded, is the color code compatible with other EFB color conventions? (That is, are there any direct conflicts in color meaning between the EFB system and the chart application?)

d) If own-aircraft position is displayed, is it shown only on charts that are drawn to scale ("geo-referenced")?
- Is the displayed position accurate to within the scale of the chart and does it remain accurate as the crew zooms?

e) If the chart application allows the crew to change between north-up and heading/track-up orientation, is the current orientation clear from the display behavior and/or a mode indicator?
- If crews became confused about the display orientation, could significant errors result?

f) Are charts printed from an electronic chart application as usable as the original paper documents?

5.1.4 Interacting with Charts

a) Can crews use the electronic charts as well as they can use paper charts?
- Can crews find and read specific detailed information (e.g., a radio frequency) on the electronic charts quickly (using zooming and panning as needed)?
- Can crews use the electronic charts to orient themselves and track their progress as they fly the procedure (using zooming and panning as needed)?
- Is there significant workload associated with configuring the electronic charts while flying the procedure (e.g., zooming/panning or other display customization)? Is display reconfiguration necessary often?

b) If de-cluttering is supported, can the crew easily switch between a de-cluttered and normal (not de-cluttered) display?
- Is there a clear indication if and when any safety-related display elements are suppressed?

C. Additional FSDO/PI Operational Evaluation Questions

This section contains additional questions that are appropriate during an evaluation by the FSDO/PI. Questions that are specific to the FSDO/PI are generally related to documentation and long-term use of the EFB (e.g., during the 6-month operational evaluation). Questions in Section C are not appropriate for the AEG/ASI.

1. GENERAL EFB SYSTEM

1.1 General Considerations

No additional questions for FSDO/PI.

1.2 Physical Placement

No additional questions for FSDO/PI.

1.3 Training/Procedures Considerations

1.3.2 EFB Documentation and Policy

See also 1.3.2 in Section A (p. 7)

a) Does the air carrier have an explicit policy that addresses the use of the EFB in line operations?

— Is the policy easy to understand and follow?

— Is it distributed to applicable personnel?

— Does the policy adequately address each specific EFB application?

b) Did the operator incorporate EFB information from the manufacturer into its existing operating documents? (See also Appendix 1, "EFB Operational Approval Process")

1.3.3 EFB Training

See also 1.3.3 in Section B (p. 15)

a) Does the carrier's initial EFB training include evaluation of knowledge and skill requirements?

— Does the training simulate key tasks?

b) Does the carrier's recurrent or continuing qualification training include evaluations of proficiency with the EFB during all appropriate evaluation gates?

1.3.5 User Feedback

a) Does the 6-month operational evaluation phase require that pilots and other users of the EFB provide post-flight evaluations?

— Is there a formal process for gathering feedback about the EFB and its support? Will feedback from this process be sent to the equipment manufacturer?

b) Does the operator provide input from personnel responsible for maintenance and data base management during the 6-month operational evaluation period?

1.4 Software Considerations

No additional questions for FSDO/PI.

1.5 Hardware Considerations

1.5.1 Display

See also 1.5.1 in Section A (p. 10)

Does the display continue to be usable after prolonged use in the flight deck environment?
- For example, can the device be damaged under normal usage?

2. ELECTRONIC DOCUMENTS

No additional questions for FSDO/PI.

3. ELECTRONIC CHECKLIST SYSTEMS (ECL)

No additional questions for FSDO/PI.

4. FLIGHT PERFORMANCE CALCULATIONS

No additional questions for FSDO/PI.

5. ELECTRONIC CHARTS

No additional questions for FSDO/PI.

Appendix E: Line Operations Evaluation Job Aid

This tool provides a starting point for EFB line-operations evaluations by the FAA inspector and operator. The questions are designed to collect a structured set of observations about use of the EFB before or during the 6-month operational evaluation. Use of this tool can be customized as appropriate for the situation.

The questions below encompass the operations and safety related functions that a Principal Inspector (PI) would normally evaluate. System complexity, software applications, mounting method, or type of in-flight use may dictate more in-depth evaluations. In cases where the FAA team finds that a system shows weaknesses or limitations, mitigations should be developed in consultation with the applicant.

In some cases an EFB may add to the complexity of flight operations. The key questions to be answered are:

1) Can the flight be conducted as safely with an EFB as with the methods/products it is intended to replace?
2) Does the EFB add an <u>unacceptable</u> level of complexity for any critical activity or phase of flight?

In order to answer these high-level questions, it is helpful to consider more specific aspects of EFB usage, which are covered in Sections II through V below. Space is also provided in Section I to record general notes about the system and the evaluation.

I. Evaluator Notes. (e.g., system description, flight conditions)

II. Overview. The main aspects to be assessed are encompassed by the following questions:

1. Was *training* adequate to ensure that the pilot(s) could perform in a safe and efficient manner?_____

    - Were individual pilot knowledge and skills adequate to allow *normal* coordinated cockpit activities?_____

    - Was pilot knowledge regarding observed software applications adequate?_____

2. Are adequate *procedures* in place to ensure that the EFB is integrated into the operator's system (e.g., normal and abnormal/emergency operations and maintenance functions)?_____

3. Were there any system hardware or software inadequacies during the flight that created a significant problem, particularly in a critical phase of flight?_____

    - Could the pilot(s) recover from usage errors without undue distraction or discussions?

    - Were usage errors frequent? Describe:_____

4. Was the workload required for completing a task with the EFB equal to or less than the workload for completing the task with the conventional method?_____

    - If no, specify phase of flight and task for any marginal or unacceptable increases in workload_____

    - Is the overall EFB workload acceptable?_____

III. General

1. Hardware (physical dimensions, input devices, display quality, arrangement/accessibility of controls, etc.):

    - Was each pilot able to use the cursor, track ball, touch screen, etc. for menu and functionality without frequent errors?___

    - Did any environmental factors (e.g., turbulence, cold weather, vibration) impact use of the EFB?_____

    - Were there significant limitations viewing the display (e.g., at off-axis angles, or under different lighting conditions)?

    - Was a screen or display ever misinterpreted because of viewing limitations?_____

    - Can any controls be activated or deactivated inadvertently? If errors are made, was it clear what the problem was and how to fix it?_____

    - Is screen brightness or background cockpit lighting an issue (e.g., at nighttime)?_____

    - Did the pilot(s) ensure proper installation and security (i.e. between flights, etc.) of EFB per SOP?_____

    - Are procedures for physical installation and security adequate?_____

- Does the display continue to be usable after prolonged use in the flight deck environment (if applicable)? _____
2. Did normal functions (e.g. shut down, start up, etc.) require undue pilot attention or concern? _____
3. Were procedures adequate for identifying currency of EFB data? _____
4. Could the pilot(s) easily find and use required items and functions? _____
5. Did the pilot(s) have difficulty understanding abbreviations or icons? _____
6. If multiple applications are supported, was there more than one critical application or function needed on an EFB at the same time and could the pilot(s) easily switch between critical applications? _____
7. Where critical items are approved (e.g., abnormal or emergency checklists) is their use at least equal to or better than previously approved methods? _____
8. Did the pilot(s) take too much time to complete normal tasks? _____
9. If audio is available, did it cause any pilot distraction? _____

IV. Electronic Charts, Documents, and Checklists

1. Were all necessary documents (including charts, checklists, and manuals) found, identified, and easily viewed by the pilot(s) without undue distraction? _____
2. Was information contained in electronic charts, documents, and checklists complete, equal in quality to previously provided products, and easily accessible and understandable? _____
3. Was pilot knowledge of chart/document/checklist selection and viewing adequate? _____
4. Could the pilot(s) easily rearrange content on the screen to meet needs (e.g., by zooming, panning, or otherwise customizing the view)? _____
5. If printers are used, are printouts acceptable? _____
6. Were all required charts, documents, and checklists available during flight? _____
7. Was legibility and accessibility of information on charts, documents, and checklists acceptable? _____
8. Are all aspects of functionality (i.e. pan, zoom, scroll, etc.) adequate and intuitive during flight? _____
9. For electronic charts:
   - Did the pilot(s) exhibit adequate knowledge of EFB functions to efficiently brief and fly required procedures? _____
   - Were both pilots able to monitor necessary electronic chart displays during critical phases of flight? _____
   - Did the system allow quick entry of updates for last minute changes (e.g., flight plan/runway changes)? _____
10. For electronic checklists, was there difficulty in tracking completed items? _____

V. Flight Performance Data/Calculations

1. Could the pilot(s) interpret and use flight performance data/calculations efficiently and accurately? _____
2. Did the system allow quick entry of updates for last minute changes (e.g., flight plan/runway changes)? _____

VI. General Conclusions

Were any unique safety issues or events caused or exacerbated by using the EFB during this evaluation? _____
Can the flight be conducted as safely with an EFB as with the methods/products it is intended to replace? _____
Does the EFB add an <u>unacceptable</u> level of complexity for any critical activity or phase of flight? _____

Appendix F: Tool Information Tables

Table F-1. EFB Human Factors Design Review Checklist

Scope and Description	This tool is used for a detailed design review of the user interface. The tool contains nearly 200 specific items. It is applicable to all EFBs. There are specific items for four common applications (electronic documents, electronic checklists, flight performance calculations, and electronic charts).
User	This tool is best suited for use by a manufacturer or designer, particularly an applications developer. Human factors expertise is not required to use this tool effectively.
Process	A single evaluator can complete this evaluation at his or her own pace, although the evaluation could also be done in pieces, by more than one evaluator. Every item on the tool should be considered, one at a time.
Documentation	The evaluator should record, in their judgment, whether the EFB is in compliance with each item. If there is an item where this is not the case, the evaluator should make an initial assessment about the criticality of the problem (e.g., Must the problem be resolved in order to obtain system approval? If not, then what will be the impact of either addressing the problem or not addressing the problem?). Specific examples of non-compliance should be recorded for discussion with system designers/developers.
Time and Resources	The evaluation can be done in an office environment. It takes approximately one-half to one full day for someone who is familiar with its items to complete an evaluation with this tool. A half-day evaluation may be enough for simple EFB systems. A whole day is required for an evaluation of an EFB with multiple functions.
Benefits	The tool is useful for catching straightforward and specific design problems (e.g., choice of font) quickly.
Limitations	The tool is best used early in the system development. The specificity of the items makes this tool less useful for catching problems that are subtler or more global, such as navigation philosophy. This tool does not address operational use of system (e.g., training and procedures).

Table F-2. EFB User-Interface Assessment Tool

Scope and Description	This tool is used for a high-level analytical ("desk top") evaluation of the user-interface, at any level of system maturity.
	The tool provides a short (2.5 page) list of EFB user interface topics to consider. The topics cover a wide range of user interface characteristics.
	This tool is applicable to all EFBs. There are specific items for four common applications (electronic documents, electronic checklists, flight performance calculations, and electronic charts).
	Because the capabilities and designs of EFBs vary from system to system, there is some overlap between the topics. This helps to ensure that all aspects of the user interface are considered at some point during the evaluation.
User	This tool may be used as a reference by anyone evaluating an EFB, including civil authorities, system manufacturers, aircraft operators, and applications/system developers. Human factors expertise is not required to use this tool effectively.
Process	The evaluation can be done by an individual, or in small teams of two or three evaluators. The evaluator(s) go through the items for each topic, commenting on each one. It is okay to skip around the list, but make sure that every item is discussed.
	The tool can be used at different stages of system maturity, to help track system improvements.
	Evaluations of the same system by multiple evaluation teams can be conducted and then synthesized to gain a deeper level of understanding. An investment of a few days can help to uncover subtle global issues, and is well worth the additional effort.
Documentation	For each item in the tool, the evaluator(s) should record any issues, and provide supporting examples from the EFB. If she or he chooses, the evaluator can also provide preliminary assessments of problem severity.
	A report containing a prioritized list of issues should be generated. This list can be tracked across evaluations to track progress on usability issues. For an example report, see Chandra, Yeh, and Riley, 2004.
Time and Resources	A single evaluation using this tool takes approximately one hour.
	The evaluation can be conducted in either an office or aircraft environment.
	Multiple evaluation sessions (i.e., more evaluators and time) are needed to get the most benefit from the tool.
Benefits	This tool can help to uncover "big" issues (e.g., potential for confusion) quickly.
	With data synthesis across multiple evaluation sessions, the tool can help to uncover subtle structural problems in the user interface.
	The tool is useful for validating the EFB system design concept at an early stage.
Limitations	Data collected from this tool are subjective and qualitative so the impact of the issues is difficult to quantify and document.
	This tool does not address operational use of system (e.g., training and procedures).

Table F-3. Guide for Developing Simulator and Validation Flight Scenarios

Scope and Description	This tool contains sample event-based scenarios that help the evaluator to construct EFB validation scenarios for flight or simulator tests. The scenarios address all aspects of the EFB, including installation, hardware, user interface, and the operational use of the EFB system, especially in unusual operating conditions. The tool only provides examples, which need to be tailored for the particular EFB system and usage being evaluated.
User	This tool is best suited for use by aircraft manufacturers or operators because of the need for a simulator or aircraft platform for the tests. Regulatory authorities may request a simulator or validation flight during the approval process, which could also be designed with the aid of this tool. Human factors expertise is beneficial in designing the scenarios, in order to plan for the data collection and analysis.
Process	It is the evaluator's responsibility to review the tool, and then propose tests appropriate for the applicant. Each operator's proposed EFB functionality and software will vary and scenarios must be customized for the particular situation. The tests should be negotiated between the inspector and applicant to minimize cost and risk. If the EFB supports applications that are not mentioned in the tool, the evaluator should develop similar tests for these other applications. If the operator prohibits use of the EFB under certain conditions (e.g., takeoff or landing), those scenarios need not be considered during the evaluation. In addition to scenarios, consideration should be given to identify what data will be collected, and what analyses to conduct. The EFB User-Interface Assessment Tool and/or the Line Operations Evaluation tool could be used to collect data. After the scenarios are performed, any problems encountered will need to be discussed and resolved between the inspector and applicant.
Documentation	The data collection and analysis should be documented in a report.
Time and Resources	Costs for flight and/or simulator tests are high in comparison to tests in an office environment. However, the EFB validation test could be combined with other necessary tests. The tests could be conducted as part of the operator's field evaluation of the EFB, or, if the operator has approved line operational simulator scenarios, the EFB could be integrated into these existing scenarios to provide a basis for evaluation. The incremental costs of flight/simulator may be acceptable, especially when testing sophisticated, or highly complex EFB systems (e.g., EFBs that are integrated with aircraft systems).
Benefits	Flight and/or simulator tests can validate overall system use under unusual operating conditions (e.g., low-visibility or emergency operations). Flight and/or simulator tests can also provide quantitative data about EFB system use. They can also be used throughout EFB system development (from concept to mature design).
Limitations	Flight and/or simulator tests may not be worth the cost for simple or evolutionary EFB systems.

Table F-4. Operational Evaluation Questions

Scope and Description	This tool is used for a comprehensive evaluation of an initial EFB system installation. It is designed to promote a thoughtful structured exploration and review of the EFB system from a human factors perspective. The whole system is considered, including system design, installation, training, operational policies, and procedures.
	The tool consists of an 18-page list of questions. Approximately 50 topics are addressed. Many of the questions are open-ended. Several also have sub-items, which are questions or considerations that clarify and expand upon the primary question. Some of the questions may not apply to a particular system because the questions address a wide variety of operators/equipment.
	There are specific questions for four common applications (electronic documents, electronic checklists, flight performance calculations, and electronic charts).
User	The intended user for this tool is a regulatory inspector whose responsibilities cover both the EFB system and its operational use. Ideally, the inspector should be familiar with the list of questions prior to the evaluation. The operator and manufacturer of the EFB system should be prepared to support the inspector's evaluation.
	Human factors expertise is not required to use this tool effectively. However, experience evaluating aircraft systems is assumed.
Process	Individual inspectors may use the questions however they find most comfortable. In general, they should use the list as a checklist of topics to examine.
	The evaluator will need to make judgments about which questions apply to a given applicant, and what response is acceptable. Once the inspector is familiar with the list of questions he/she will be able to quickly determine whether a question is applicable or not to the current evaluation.
	In some cases, the inspector's assessment of the system will conflict with the applicant's assessment. These conflicts should be resolved cooperatively. Mitigations should be developed among all parties (regulator, operator, and the manufacturer, if appropriate).
Documentation	The evaluator should take notes on each item that they consider from the set of questions. Any concerns or questions should be documented, along with specific examples whenever possible. FAA evaluators should discuss these with the applicant.
Time and Resources	The evaluator should allow approximately one day for the evaluation. Some portions of the evaluation (e.g., assessment of training) could be done in an office environment, but the aircraft and installation should be well understood in advance.
Benefits	This tool can be used to validate initial use of EFB system for a particular flight deck. The tool considers all aspects of EFB use, including system design, installation, training, operational policies, and procedures.
Limitations	The utility of these operational evaluation questions is dependent upon the evaluator's experience in customizing the tool for the evaluation and in making appropriate assessments regarding need for training, etc.

Table F-5. Line-Operations Evaluation Job Aid

Scope and Description	This tool helps the evaluator record observations about in-flight use of EFB systems. These records help the evaluator to assess the EFB system in normal use by line pilots.
	The tool fits on a single sheet of paper printed on both sides for ease of use.
	The questions cover general operations and safety related functions. They help the evaluator to assess (1) the overall safety of the EFB system compared with the methods/products it is intended to replace, and (2) whether the EFB adds an unacceptable level of complexity for any critical activity or phase of flight.
User	This tool is designed for use by the inspector during an operational evaluation period of several months. An operator can also use it effectively for internal evaluations prior to the formal operational evaluation. Human factors expertise is not required to use this tool effectively.
Process	This tool is intended for use in evaluating EFB system use over multiple observations. The evaluator will fill out the tool either during, or, more likely, after an observation flight, to record his/her impressions of EFB use by the pilots. The evaluator should be familiar with the questions in advance of the flight so that he/she can observe crew use of the EFB from an appropriate point of view.
	Use of this tool can be customized as appropriate for the situation. Some questions may not apply to a particular EFB system. An inspector should consider the applications and operational use planned for the EFB in customizing the tool.
	An operator could customize this tool to obtain more quantitative data for internal assessments. In particular, operators could ask for a numerical response rating (e.g., low to high frequency of event) to obtain a better understanding of how to improve the system for efficiency, as opposed to just making it acceptable to an authority. The distribution of the scores could identify issues where there is large variability between flight crews, which could indicate a need for improved training, or other mitigations.
Documentation	A form should be filed for each observation flight. All the observations for a particular evaluation should be reviewed in aggregate to discover whether there are any patterns to problems or issues that arise in normal use. No *single* observation of EFB use should determine whether the system is acceptable or not
Time and Resources	The evaluator should allow approximately 15 minutes during, or after each observation flight to fill out the tool.
	Operators who add numerical response ratings will need to enter the responses into a spreadsheet to see the distribution of scores, and to compute statistics on the data.
Benefits	This tool helps to validate overall EFB system use during normal operations at a relatively low cost.
	The tool can help to uncover training and procedural issues. The tool can also uncover variability in performance between pilots.
Limitations	This tool does not evaluate the user interface in detail. Also, in its simplest form, the tool does not collect quantitative data.

www.ingramcontent.com/pod-product-compliance
Lightning Source LLC
Chambersburg PA
CBHW081853170526
45167CB00007B/3004